Troubleshooting and Repairing Commercial Electrical Equipment

About the Author

David Herres is the owner and operator of a residential and commercial construction company. He obtained a Journeyman Electrician's License in 1975, and has certificates in welding and wetland delineation, along with experience with elevators. Beginning in 2001, Mr. Herres has focused primarily on electrical work, upgrading his license to Master status. The author of an article for *Electrical Construction & Maintenance* on complementary writing methods, he has written nearly 70 articles on electrical and telecom topics. Mr. Herres resides in Clarksville, New Hampshire.

Troubleshooting and Repairing Commercial Electrical Equipment

David Herres

New York   Chicago San Francisco
Lisbon London Madrid Mexico City
Milan New Delhi San Juan
Seoul Singapore Sydney Toronto

Cataloging-in-Publication Data is on file with the Library of Congress.

McGraw-Hill Education books are available at special quantity discounts to use as premiums and sales promotions, or for use in corporate training programs. To contact a representative please e-mail us at bulksales@mcgraw-hill.com.

Troubleshooting and Repairing Commercial Electrical Equipment

1 2 3 4 5 6 7 8 9 0   DOC/DOC   1 2 0 9 8 7 6 5 4 3

ISBN   978-0-07-1810302
MHID   0-07-1810307

The pages within this book were printed on acid-free paper.

Sponsoring Editor	**Copy Editor**	**Composition**
Judy Bass	Susan Fox Greenberg	Cenveo Publisher Services
Acquisitions Coordinator	**Proofreader**	**Art Director, Cover**
Amy Stonebraker	Megha Saini, Cenveo Publisher Services	Jeff Weeks
Editorial Supervisor		
David E. Fogarty	**Indexer**	
	ARC Films Inc.	
Project Manager		
Vastavikta Sharma, Cenveo® Publisher Services	**Production Supervisor**	
	Pamela A. Pelton	

Contents

Preface

This book is offered up in the hope that it will provide a gateway to greater knowledge and expertise in the fascinating field of electronics, and specifically that it will provide the reader with intellectual tools needed to successfully diagnose and repair electrical equipment. Our primary focus is commercial and industrial machinery, but overlap is inevitable, so we find ourselves considering everything from three-way switches in a residence to the International Space Station in the sky above us.

A certain amount of knowledge is presupposed. It is assumed that the reader, like most people in today's wired world, knows the difference between direct and alternating current, that two conductors are (usually) required to complete a circuit, and that receptacles are wired in parallel daisy-chain style while switches are placed in series with the load. If you have no idea what I am talking about, consult some of the elementary tutorials that are available online before proceeding. Otherwise, you are good to go.

As for mathematical savvy, we presuppose a knowledge of high-school algebra. You do not need a background in trigonometry or calculus to understand what is discussed in this book. To review some basics, refer to Stan Gibilisco's excellent *Everyday Math Demystified, Second Edition*, McGraw-Hill, 2013. Another helpful reference by the same author is *Electricity Demystified, Second Edition*, McGraw-Hill, 2012.

Before diving in, please be aware that there are tremendous moral and legal implications whenever you work on an electrical infrastructure or equipment that your co-workers or members of the public may approach. Every design detail and electrical termination has to be right; otherwise, a hazard is built into your work. The result can range from an inconvenient outage to a tragic inferno.

I have a lot to say about electrical hazard mitigation in my previous McGraw-Hill book, *2011 National Electrical Code Chapter By Chapter*. I have also written numerous magazine articles on electrical topics and many are archived on my website, www.electriciansparadise.com. I welcome your thoughts and comments. Write to me at David@electriciansparadise.com.

Acknowledgments

I would like to thank Judi Howcroft, renowned nature photographer, philosopher, and friend of the oppressed people of the world, for her insight and encouragement. Most of the photographs and all of the inspiration for this book are hers.

David Herres

Troubleshooting and Repairing Commercial Electrical Equipment

Troubleshooting Techniques

To a certain extent it is possible to repair a piece of electrical equipment without a clear understanding of its inner workings. You may find a failed component by visual inspection and, by replacing it, restore the machine to operation. Alternately, it is sometimes possible to perform one or more therapeutic operations such as pulling apart and reattaching every ribbon connector in order to polish the contacts or resoldering all terminations in a printed circuit board in hope of eliminating an offending cold joint. In the absence of a true understanding of the equipment, these approaches may be valid, but the success rate is low. It is better to strive for knowledge of the structure and operating modes of the machine.

Years ago, I was called upon to repair a two-way radio base station. At the time, I had only an impressionistic knowledge of such circuitry. The unit would power up, but it would not transmit or receive, and there was no background noise emanating from the speaker with the volume turned all the way up.

I removed an access panel and it was plain to see that a circuit board had become detached from its mount so some electrical terminations could ground out against the inside of the enclosure. I remounted the circuit board, plugged in the base station, and found that it was operational.

Later, I learned in a roundabout way that the piece had been dropped and that is when the trouble started. That would have provided an important clue. Many times such information is not forthcoming; however, any background history provided by the operator is of great value.

The visual inspection method worked in that case, but it cannot be depended upon. Often components go bad without any noticeable change in appearance, and more advanced sleuthing techniques become necessary. Moreover, circuit defects such as degraded terminations or shorted circuit board traces (caused by an errant metal filing or conductive dirt accumulation) may not show up visually.

The visual inspection method, while appropriate as a preliminary measure, is accompanied by some pitfalls.

- First, especially in modern equipment, it may not be immediately evident how to open it. The quality of plastic material has improved greatly in recent years, and is now used for equipment enclosures and mechanical parts even for very advanced and high-quality products such as digital cameras and submersible pumps. Plastic enclosures typically have some odd ways of snapping together, and may be reluctant to come apart. Obviously, it is never good to use excessive force. Specialized tools are available. In addition, expedients such as a plastic guitar pick or an old credit card to slide into narrow gaps are effective. A service manual may provide the key, and YouTube videos pertaining to all sorts of equipment are abundant.

- A second problem in relying on visual inspection is that an obviously burnt component may not be the original culprit, but instead a consequence of some other failure mode that caused excessive current to be delivered to the site of destruction. Furthermore, in opening up it may have functioned as a fuse, preventing additional damage to the circuit. Replacing such a failed component may jeopardize other elements, making for a more difficult and expensive repair. Therefore, we see that operating from a more enlightened perspective is usually the way to go.

It is possible to damage the very equipment you are trying to fix. Some solid-state components, especially complementary metal-oxide semiconductor (CMOS) transistors (we will discuss them later), operate at exceptionally low supply voltages and input signal levels, and they are quite sensitive to static charges that may be impressed upon them in handling. A copper bracelet equipped with a connecting wire to a known good ground is available for electronics technicians and it serves to drain static buildup from the human body. At the very least, circuit boards should be handled at the edges only. Soldering irons and other conductive tools must also be protected against static charge so that they do not transfer it to the electronic components.

> If a chunk of concrete or similar material has become lodged in an empty conduit, chuck up an appropriate length of fish tape in a drill and use it to break through the offending item. Then use your shop vac to remove the debris. As a corollary, save short lengths of broken fish tape as they are handy in many situations.

Electronic equipment must be protected from damage by trauma while it is in your care. The work surface should be nonconductive material, neither grounded nor ungrounded. If a computer monitor or TV is to be laid screen down, it should be placed on a soft folded blanket that will protect sensitive surfaces from scratches and help keep them clean. If bench grinders and drill presses are to be used, they should be located in a separate room to ensure that conductive filings do not accumulate.

A strictly visual approach may be helpful, but it cannot be relied upon. A general knowledge of electronic theory plus troubleshooting techniques and practical repair methods, as well as knowledge specific to the equipment under consideration is necessary. It is hoped that this book will open some doors. The mistake many aspiring technicians make is thinking they are going to fix an ailing electronic device in 15 minutes. If the machine is of even moderate size and complexity, it will take longer. You need to approach the problem in a calm and orderly fashion, and not expect to find the defect immediately. The right diagnostic and repair tools are essential and the foremost tool is knowledge. Fortunately, this commodity is readily accessible. A vast amount of equipment documentation is free on the Internet and, more intensively, in print. A textbook may seem expensive, but one good idea gained from it will easily pay for the volume.

A further issue has come up in recent years. Increasing miniaturization and use of proprietary microprocessors has meant that a lot of equipment that is around these days is said to be unserviceable. To a limited but not absolute extent, this is true, but that is far from the whole story. With the right equipment, as we shall see later, very small microprocessors with many pins may be desoldered and replacements installed. It is feasible to repair printed circuit boards and, once a few simple principles are understood, the soldering techniques are not too difficult. Consumer appliances may be difficult to disassemble and diagnose, but the necessary information is out there and consistent success in this area is achievable.

Entering the Field

Before proceeding, we should make some other observations. This book has for a premise the notion that the fields of electricians and electronics technicians, while remaining separate and distinct groups of workers, are beginning to merge. This is because of several recent trends. Economic dislocations, not to say turmoil on a worldwide level, have meant at times a marked reduction in commerce and decreased demand for services. Knowledge-based professions including electricians have been partially protected from the worst effects of this slowdown. While there has been diminished demand for new construction, maintenance and repair have remained strong.

Many electricians have been able to redefine their way into low-voltage work, by which is meant telecom, alarm, and data. (Low voltage is something of a misnomer because at times the voltage and power levels are sufficiently high to be hazardous.)

Regardless, such a shift provides access to much more work, making the profession relatively recession-proof. For example, new commercial buildings in most jurisdictions are required to have centralized fire alarm systems. This goes far beyond simple smoke detectors even if wired in concert to go off simultaneously. Such fire alarm systems are costly and complex but very reliable in reporting fires, even if at times prone to false alarms. Many commercial venues such as large restaurants and hotels have robust fire alarm systems, which are great life and property savers, but nuisance alarming can be disruptive to put it mildly. Such establishments rarely have an on-site fire alarm technician, so it falls upon electricians to perform routine maintenance and, in case of malfunction, to decide when to call in a fire alarm specialist.

Some states, municipalities, and other jurisdictions require licensing for fire alarm designers and installers. However, the many ways these systems interact with sprinklers, elevators, air-conditioning and ventilation, combustible gas flow, telephone, Ethernet, and other systems, to say nothing of the electrical supply, make precise boundaries difficult to delineate. Usually the electrician is called upon to make quick, critical decisions regarding rapidly unfolding events.

Here again, knowledge is the key to successful resolution in these matters. In Chap. 12, we will take a closer look at fire alarm systems. The point for now is that they provide an opportunity for electricians looking to expand into related areas. When the fire alarm system requires attention, we enter a troubleshooting mode. The techniques are similar for all types of systems and equipment.

In troubleshooting, the first step, as mentioned previously, is to interview the operator and gather relevant information. It is especially important to determine if the problem developed gradually or appeared suddenly. Indeed, has the machine ever functioned properly? If not, there may be a manufacturing or programming defect as opposed to failure of a discrete component. The operator should also be asked if failure was accompanied by any unusual sounds, sparks, flashes of light emanating from within, dimming of nearby lighting that could indicate an excessive current drain, outside evidence of a voltage surge, burning odor or sensation of overheating, etc. Ask the operator for any thoughts on the cause of the malfunction. It is amazing how often an operator, having come to know the machine over a period of time, can have a sense of the nature of the problem. The initial interview may prove helpful in retrospect, even if definitive information does not appear forthcoming at first.

Thus far, we have been speaking of the defective unit in terms of a small machine, something that could be easily moved to the repair shop if extensive work is needed. However, the same comments apply to any size equipment from the largest machine ever built, the North American electrical power grid, to microscopic motors that are now being constructed at the very frontier of nanotechnology.

Complex machinery, such as the huge particle accelerators now in operation and straddling international borders, are equipped with a lot of instrumentation and in many ways have self-diagnostic capabilities. Commonly encountered electrical equipment may have a power light and a temperature gauge, or much more extensive instrumentation. An elevator or fire alarm system will have a control panel with alphanumeric display intended as a human interface. Much smaller equipment, such as a household microwave oven, will have a readout that displays time and operational details, and may provide a seemingly cryptic error code in the event of failure. The error code (something like E8) may be referenced in the operator's manual. If not, you can type the error code with make and model into a search engine and instantly come up with a diagnosis and suggested repair procedure.

Let us say, for now, that a machine has been examined visually and no obvious defect has been found. What next? Based on information provided by the operator, you can attempt to power up the machine. If it runs on other than 120 volts and you do not have the proper volt and ampere branch circuit and receptacle in the repair shop, you will have to run a line from the nearest panel. It is essential to verify that you have the correct size wire and overcurrent protection. Consult the National Electrical Code (NEC) for sizing and branch-circuit construction details. If the machine is relocated to the repair shop and it contains a motor or other device where the order of the phases is critical, that becomes a consideration and must be dealt with.

When wiring a three-phase motor, it is common practice to find correct phase rotation by trial and error, observing phase rotation and switching wires at any convenient location anywhere upstream provided other motors are not affected, as in a group installation. (To reverse rotation in a three-phase motor, any two of the three wires are reversed.) However, it must be noted that some equipment, notably some types of motor-driven pumps, will be destroyed instantly if run in the wrong direction, at least to the extent that the seals will be damaged and need to be replaced. Therefore, the trial and error method is not always a good choice.

A better approach is to use a three-phase motor rotation tester, which costs about $100 and comes with an instruction manual. Alternately, you can use a small, fractional-horsepower three-phase motor to check hookups at various locations by observing rotation. (You will find a more complete discussion of motor troubleshooting and installation in the next chapter).

After doing the initial hookup or beginning any repair procedure, I always check the enclosure to make sure it has not become energized with respect to ground. Check again after the repair has been completed. A ground fault inside, perhaps caused by a chafing wire in conjunction with a missing equipment ground, can make the chassis and enclosure hot, resulting in shock hazard. A good way to check this is by using an inexpensive neon test light, available from your electrical distributor or any good hardware store. This is a great instrument for checking the presence or absence of voltage and approximate voltage level (120 or 240 and up to 600) and it can be used to distinguish hot from neutral. It has very high impedance because the neon bulb draws little current and there is also a resistor in series with the bulb. After preliminary measurements, you will want to set aside the neon test light and use your digital multimeter for internal work.

As for troubleshooting technique, the following comments are applicable to all sorts of equipment, but for now, we will turn to a typical premises electrical problem. This is a very simple, almost trivial, example, but it will show how the entire electrical troubleshooting process can be rationalized to ensure success without wasted effort. By this we mean analyzing the situation based on available information and proceeding in a way that makes sense, rather than drifting from test to test in a random manner in hopes of hitting on something.

Let us suppose that we are called into a commercial area where a single ceiling-mounted two-bulb fluorescent fixture is out—no flicker or dim light output. We know by experience that most likely replacing both bulbs will solve the problem. However, suppose there is a high ceiling and there is no ladder on site that will access the fixture. Before investing time in going after a ladder, it makes sense to look elsewhere first—switch, circuit breaker, etc. even though the defect is more likely a bulb or, next in line, the ballast.

> When temporarily taping materials, leave an inch of tape. Make a tail by spinning it between your fingers, so that later you will not have to dig to find the end.

From this trivial example, it is possible to work out a general principle: prioritize all troubleshooting operations with regard both to probability of success and difficulty or expense in performing the diagnostic. Now back to the defective machine.

Getting Started

We have applied power by plugging it into a receptacle or, if hardwired, energized the line. Often there is a power light. If it does not light up or the machine is completely unresponsive even though there is voltage at the input terminals, the machine is said to be dead. These problems are actually among the easiest to diagnose. Conductors and components are large and wiring is easy to trace, even without specialized knowledge of the piece of equipment and in the absence of a schematic diagram. Look for the moving parts because these are more likely to fail. Many types of laundry equipment and portable food processors found in a commercial kitchen have door interlocks. These are safety switches that prevent the machine from running if the door is not firmly closed. The switch may be a simple single-pole device in series with the incoming ungrounded conductor. Either this interlock switch is defective (open) or there is a mechanical problem in the linkage to the door or the door is not latching completely. Pressing firmly on the door may cause the machine to come alive. The problem is easily repaired by replacing the switch or repairing the linkage, door latch, or hinges. Avoid the temptation to override the switch by shunting around it to get the machine working because the switch plays an important safety role.

There are many instances of this type whereby incoming power is interrupted. They are easy to diagnose and repair. Frequently a power wire is burnt at the termination. Another repeat offender is the main power switch. In addition, there could be a timer or fuse that malfunctions. Any of these series-connected items may be checked with an ohmmeter (with power disconnected). These passive devices may usually be checked in circuit because they are dead-ended when not powered up. Specifically, there should be no load in parallel that would fool the ohmmeter. When in doubt, disconnect one lead.

Another method for checking the functionality of switches, fuses, and similar devices is to take a voltage measurement across them while powered up. A switch should read absence of voltage when in the ON position and presence of voltage when OFF. This method is in some ways the more professional approach and is useful in checking fuses in an old entrance panel or controller where visual inspection is not always definitive.

A cautionary note: Certain components such as large electrolytic capacitors that are part of a power supply may be capable of storing a significant electrical charge that can be hazardous. It is common practice, after powering down the equipment, to short them out to remove the charge before servicing, but the high flow of current may damage the component. A better method is to shunt across the terminals using a low-ohm power resistor or other appropriate load.

Many machines have an in-line fuse. If it is blown or if there is a tripped circuit breaker, it is best to look into the matter further before powering up the equipment with new overcurrent protection. A shorted component may have activated the overcurrent protection, but next time damage may be more extensive. A chafed wire or transformer with shorted primary or secondary winding, or shorted parallel-connected component such as diode or capacitor could be the problem. Ohmmeter readings will suffice to check these devices. On the other hand, fuses may open due to age or in the course of performing their function of protecting against transient overvoltage events such as line surges.

"Half splitting" is another very powerful troubleshooting technique. Most electrical equipment and systems are serial in nature, which means the successive stages are connected so that the output of one feeds the input of the next. This is not to say that there are no other parallel elements. They may be identified by examining the wiring or looking at a schematic or a block diagram. The half-splitting technique involves going to the midpoint of the circuit and performing tests. By interpreting the results, it is possible to eliminate half of the equipment or system in one operation. Then, go to the midpoint of the "bad half" and perform another test. Proceed in this fashion to quickly isolate the defective component. In choosing the location for your measurement, select a point where access is easy. An example of this process would be in a premises wiring system where power is not present in a string of outlets. Rather than checking each outlet starting at either end, go directly to the midpoint. The same basic technique is useful in audio or video equipment where the output of one stage becomes the input of the next.

Experienced technicians may use intuition even where they are unfamiliar with the type of equipment. Others will want to consult a service manual and schematic before proceeding. By way of background information, we will now embark upon a discussion of various electronic components and their schematic symbols.

In recent years, there has been considerable standardization and simplification of schematic symbols and conventions, the result being that the diagrams are easier than ever to comprehend and use. In fact, they are extraordinarily intuitive. An example is the symbol for a capacitor, which graphically depicts two electrodes in the form of parallel plates with conductors attached and separated by a nonconducting dielectric material.

Schematic diagrams display the logical arrangement of components with interconnections shown, without regard to the parts arrangement on printed circuit board or chassis. The actual layout is shown in a pictorial diagram, which may be a photo. Here the parts locations are governed by a number of considerations including economy of wiring, protection from radiofrequency (RF) interference and thermal effects from other circuits, and maintaining characteristic impedance matching to prevent harmful data reflections at high frequencies. (More about this when we discuss impedance.)

> A fish tape is spring steel, halfway between mild steel and hardened steel. If you try to bend it too sharply, it will usually break. If the end breaks off, you can make a new hooked end by annealing it with a propane torch. Heat it red hot and allow it to cool slowly. This process softens the metal so that it can be formed with needle-nose pliers.

Electronic Components

Without further ado, we shall begin a discussion of individual components and their schematic representations. Figure 1-1 is the symbol for wires crossing and connected; Fig. 1-2 is the symbol for wires crossing, not connected; and Fig. 1-3 is another symbol for wires crossing, not connected, however this symbol is no longer used, but is still seen in old diagrams.

Wires must be sized for the maximum current they will carry, taking into consideration ambient temperature, duty cycle, type of insulation, heat produced by connected load, and other factors. When replacing a wire, it is important not to go smaller than the original. Going to a larger size can create problems as well, altering the characteristic impedance in high-frequency applications thereby causing a mismatch with harmful reflections. Insulation type should not be changed, particularly in hot locations such as the leads for heating elements. Conductors in underground raceway must always be suitable for wet areas.

Figure 1-4 is a wonderfully intuitive symbol representing a fixed resistor. Resistors may be in series or in parallel with the power source or with the load. Resistance is measured in ohms and is not frequency dependent, except for unintended effects caused by incidental inductance and capacitance at high frequencies. The basic formula is:

$$E = I \times R$$

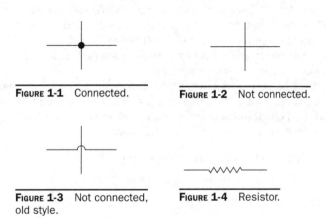

FIGURE 1-1   Connected.

FIGURE 1-2   Not connected.

FIGURE 1-3   Not connected, old style.

FIGURE 1-4   Resistor.

where E = electromotive force (in volts)

I = current (in amperes)

R = resistance (in ohms)

For a specific current, the formula transforms to:

$$I = E/R$$

If you want to choose a resistor for a specific application, use this transformation:

$$R = E/I$$

These relations plus power formulas are presented in the Ohm's law wheel (see Fig. 1-5).

One of the power formulas is:

$$P = I^2 \times R$$

where P = power in watts.

This formula is of great importance to electricians and electronics technicians. It is saying that the power in watts, for example heat dissipated by a resistor, is proportional to the square of current flowing through it in amperes times resistance in ohms.

From $E = I \times R$, we know that with any given voltage, I and R are inversely proportional. As one of these variables goes up, the other goes down. Since $P = I^2R$, the value of I is far more important than R. As I increases, P (the amount of heat dissipated intentionally or as the result of an unintentional fault) soars.

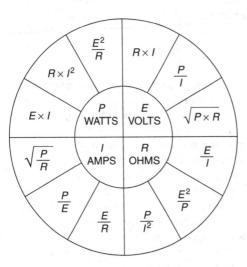

FIGURE 1-5 Ohm's law wheel.

What's the Significance?

This relationship has enormous consequences in electrical design, maintenance, and repair work. Electrical energy that disappears from a circuit reappears as heat outside the circuit, and it must be safely dissipated in order to prevent uncontrolled temperature rise accompanied by damage to equipment and possible fire hazard.

Electrical engineers speak of I^2R heat loss, and it must be considered when sizing out all sorts of electrical equipment including resistors.

Accordingly, these passive devices are rated not only in ohms for resistance, but also in watts for power handling ability. In replacing a resistor, it is important to use the same size or larger. If the original burnt out, a larger replacement may be indicated. Never go smaller.

In electrical equipment, resistors play numerous roles:

- In series with a load to limit current through the load

- In parallel with a load to limit voltage on the load

- In conjunction with a power supply to create a voltage divider

- Biasing of active components
- Bleeding of capacitors in a power supply
- Damping of LC tuned circuits
- Setting time constants in RC circuits
- Dummy loads

A defective resistor will prevent the circuit with which it is associated from working. It may take out the circuit, stage, and entire piece of equipment. Defective resistors are usually open and may or may not visually appear burnt.

There is a universal resistor color code. (See "Resistor Color Code.")

Resistor Color Code

The value in ohms of a resistor is denoted by the colored bands that surround it.

There are usually three bands, then a gap, then a fourth band. (By the position of the gap, you can tell which end is the beginning.)

The first three bands give the resistance.

The first band is the first significant figure.

The second band is the second significant figure.

The third band is the multiplier (number of zeros following the second significant figure.)

After the gap, the fourth band is the tolerance.

The colors, with numbers, are:

- Black = 0
- Brown = 1
- Red = 2
- Orange = 3
- Yellow = 4
- Green = 5
- Blue = 6
- Violet = 7
- Gray = 8
- White = 9

The tolerances are:

- Brown = 1 percent
- Red = 2 percent
- Gold = 5 percent
- Silver = 10 percent

If the fourth band is missing, the tolerance is 20 percent.

It is not necessary to memorize the color code. Just hang it above your bench.

The schematic symbol for a capacitor makes sense both functionally and structurally (Fig. 1-6). It may be thought of as two large-area metal plates, closely spaced, separated by air or other insulating material. In the real world, you would more likely see two long strips of foil with a strip of insulating material, rolled tightly into a cylinder, with leads attached. A ceramic capacitor, for high-frequency RF applications, has the two electrodes embedded near enough to one another to make a low-capacitance device. In the electrolytic capacitor, shown in Fig. 1-7, the dielectric layer is formed chemically when electrical energy is applied. This layer is very thin, making for high capacitance.

FIGURE 1-6   Capacitor.

Capacitors take many forms, the shape and material depending upon electronic and economic considerations. There are always two electrodes or sets of electrodes separated by an insulating layer that maintains the spacing between the plates and prevents them from touching and shorting out. Actually, this dielectric layer plays a far more complex and important role in every capacitor. A capacitor is similar to a resistor in the sense that it opposes the flow of current through it. (In fact, current does not actually flow "through" a capacitor. Energy is conveyed by means of an electrostatic charge on the dielectric material. However, we can talk as though there is a current flowing through the device.)

FIGURE 1-7
Electrolytic capacitor.
(*Courtesy of Judith Howcroft.*)

To understand how a capacitor works in an electronic circuit, it is necessary to know a few definitions.

Impedance is the total opposition to the flow of current. It is made up of resistance, capacitive reactance, and inductive reactance. It is measured in ohms. In a circuit, it works essentially the same as resistance and it conforms to Ohm's Law in the same way as a resistance.

Capacitive reactance is opposition to the flow of current through a capacitor or through any part of a circuit that has capacitance, either intended or unintended (called parasitic capacitance). Unlike resistance, capacitive reactance is frequency dependent. It is given by this formula:

$$X_C = 1/2\pi f C$$

where X_c = capacitive reactance
C = capacitance (in farads)
f = frequency (in hertz)

As you can see, a capacitor offers more opposition to the flow of current at lower frequencies and less opposition at high frequencies. To very high frequencies, a capacitor is invisible. To dc, a capacitor offers maximum opposition and constitutes an open circuit, except that there is a current spike at the instant that the dc is switched on or switched off, at which point it has the property of a high-frequency pulse due to high rise and fall times.

Inductive reactance is a parallel concept, with the circuit properties being a mirror image of the above. We will discuss inductive reactance in the section on coils.

It is necessary to distinguish between capacitance and capacitive reactance. Capacitance is a fixed property of the capacitor or circuit. It is given by involved formulas including area of the plates (greater area means more capacitance) and distance between plates (closer proximity equates to more capacitance). Another variable is the dielectric constant of the material between the plates.

It is not necessary for the electronics technician to know or use the formulas for capacitance. The manufacturer will provide a capacitor having the desired capacitance, which will be printed on the device or indicated by color code.

When tapping out small threaded holes like in a wall box, place your tap in a cordless drill and feed it in slowly. This procedure gives better control and is much quicker than using a wrench or socket to hold the tap. Do not forget to oil the tap, preferably with thread cutting lubricant.

The other important parameter of a capacitor is working voltage (WV). If a capacitor is deployed in a circuit where it is subject to more than its WV, the dielectric layer may be punctured by arcing electrical energy and the capacitor permanently ruined.

Capacitance is measured in farads, but capacitors we encounter in the course of most work are rated at a small fraction of a farad. Microfarad and picofarad are the usual metrics.

Frequency Is All-Important

In contrast to capacitance, capacitive reactance, as we have seen, varies with frequency and must be calculated based on the capacitance of the device and the frequency of the voltage (hence current) involved. Therefore, the technician must know and be able to use the formula for capacitive reactance so that it can be figured on a case-by-case basis.

Because of its property of exhibiting different impedances at different frequencies, capacitors have numerous real-life applications. An example is their ability to separate different frequencies including dc.

A common usage involves network-powered communications systems. The power may be supplied at close to 60 volts. These two vastly different voltage and frequency levels can coexist within the same cable with no problem, and they can be separated by means of a capacitor network or a resonant circuit which, as we shall see later, involves use of a capacitor and an inductor in series or in parallel.

An interesting device is the electrolytic capacitor, which is very often seen in power supplies to smooth out and remove ripples from dc after it has been produced by rectifying ac. Another application is in motors where it may be switched online while the motor is starting.

Many electrolytic capacitors are easily recognized by their comparatively large, tank-like appearance. They often have lugs for push-on spade connectors and are marked with capacitance and working voltage. The electrolytic capacitor is used where comparatively high levels of capacitance are required.

Not all electrolytic capacitors have that appearance. Tantalum capacitors are a subspecies. They are very reliable, but more expensive, and because of their small size, are well suited for cell phones and similar small portable equipment.

Electrolytic capacitors are manufactured in an unconventional way. One of the plates is made of an ionic conducting liquid. When voltage is first applied, this plate manufactures a dielectric layer that is very thin. Since the plates are therefore much closer than in other types of capacitors, a much higher capacitance is possible in a relatively compact package.

Electrolytic capacitors, because of high capacitance and high-power applications, are prone to failure. They can become shorted or open, or fail to meet capacitance specifications. They may become defective if stored on the shelf for a long period of time, especially under hot or humid conditions. Sometimes these devices can be restored by applying a steady dc voltage. The good news is that they are easy to test and replace.

Electrolytic capacitors can be tested with an ohmmeter, although in an unconventional way. Connect the ohmmeter set on a medium to high range to the two terminals. Depending upon whether the capacitor is charged and which way the ohmmeter probes are connected, the digital readout or analog needle will either climb or descend in a uniform and stately fashion or it will remain unchanged. Electronics technicians call this meter response "counting." If it remains unchanged, reverse the leads and the meter will commence counting, indicating that the electrolytic capacitor is good. This odd behavior is because the ohmmeter has a battery that puts voltage on the capacitor leads in an attempt to take resistance readings. In the case of an electrolytic capacitor, however, the meter will either charge or discharge the capacitor.

This response indicates that the capacitor is good—not shorted or open—and has capacitance. However, the test is not definitive. Under dynamic circuit conditions, the device may not perform adequately.

A defective electrolytic capacitor may appear burnt, swollen, or distorted, or it may be leaking electrolyte. Such a unit must be replaced, but watch out for other bad components that may be putting too much voltage on it.

Electrolytic capacitors are everywhere—in all size motors, computers, power supplies, TVs, electrical distribution equipment, and so on. Replacement capacitors must have the same or greater working voltage. In some applications, the exact capacitance is not critical. It is usually acceptable to go 10 percent higher.

To summarize, capacitors, especially electrolytic, are frequent culprits in many equipment failure events. For most applications, unless it is a power factor device on an industrial scale, the replacement cost is modest.

The coil schematic illustrates the structure of this ubiquitous device (Fig. 1-8). Wire, usually copper, is wound in helical fashion, often around an iron core. Leads at either end provide circuit connections. The device is simple, but the underlying concept, inductance, is a large topic, involving some fundamental and mysterious properties of the universe.

FIGURE 1-8   Coil.

Actually, any conductor has inductance, whether it is an ionized channel blasted across the sky by a lightning bolt or a nerve cell axon within your brain. However, usually we are talking about a segment of copper wire. Winding it to form a coil multiplies the inductance, and if these windings encircle a core made of a magnetically permeable material such as soft iron, the effect is increased dramatically. All materials are to some extent permeable to magnetic flux, and a vacuum is as well. (Permeability in a magnetic circuit is like conductivity, the reciprocal of resistance, in an electrical circuit. Magnetic flux is like electrical current.)

Any coil or conductor has inductance, and this is a fixed value, whether the coil is on the warehouse shelf or connected to a live circuit.

There are some complex formulas used for calculating this value. It is not necessary to perform these calculations because we are not in the business of designing and building inductors, or designing and building long electrical or telecom distribution lines. The inductance will be marked on the device by the manufacturer, and we may take this as a given when ordering replacement parts. Remember that inductance is a property of the device and remains unchanged under normal conditions, that is, regardless of frequency, voltage, current, or any other circuit parameters.

In contrast, inductive reactance is a property of the device that is dependent upon both the inherent inductance of the coil or conductor, and how it is deployed in the circuit— specifically the frequency of the current passing through the conductor. Inductive reactance, like the capacitive reactance that we discussed earlier, is a measure of opposition to the flow of current. It is likewise measured in ohms, and conforms to Ohm's law. The way inductive reactance differs from resistance is that it is frequency dependent.

Inductance may be said to be a mirror image of capacitance, and like any reflected image, certain elements of the reflection are reversed. Specifically, in a capacitor or capacitive load, the capacitive reactance is greater at lower frequencies (maximum at dc, 0 Hz) and less at higher frequencies, whereas in a coil or inductive load, the inductive reactance is just the opposite. It is greatest at high frequencies and less at low frequencies. At dc, or 0 Hz, there is no inductive reactance except at the instant when power is applied and when it is disconnected, at which times the current behaves like a high frequency, with fast rise time and fall time of the waveform. In this connection, it is interesting to note that when power is abruptly removed from a circuit that contains a high-inductance coil in series with the load, there can be a quite significant voltage spike. The magnitude of this pulse is dependent upon the preexisting voltage level, the amount of inductance in the circuit, any series or parallel impedance, and especially the speed of the switching action. The phenomenon is known as "voltage kick" and it can be unintended and harmful, or it can be part of the equipment design. An example is the old mechanical ignition for a gasoline engine, with rapid switching provided by the points in conjunction with a spark coil.

Regularly inspect the handles of insulated tools for deterioration, which can be hazardous. An almost invisible crack can hold moisture and conducting grease. Heat-shrink tubing installed on the handles of pliers will make them like new but insulated handles should never be trusted for high voltage or when you are on a wet surface or otherwise grounded.

Inductance is directly associated with the interaction between electricity and magnetism. When current flows through a conductor, a magnetic field is established inside the conductor and in the space surrounding the conductor. The strength of the magnetic field, of course, diminishes with distance from the conductor.

When voltage is applied across the conductor, accompanied by current flow through it, the magnetic field is established, and the energy has to come from somewhere. Accordingly, there is an inevitable reduction in current flowing through the conductor. We may consider that the magnetic circuit is borrowing energy from the electrical circuit. When the voltage is removed or even reduced, it is payback time! As the magnetic field collapses, energy is fed back into the electrical circuit.

There are two kinds of inductance: self-inductance and mutual inductance. It is the same phenomenon, just a different arrangement of the energy flow. Self-inductance always occurs when there is flow of electrical current. Mutual inductance takes place when a second coil, often wound around the same core, is brought in close proximity to the first coil. Mutual inductance is also relevant when two conductors are run parallel to one another for an appreciable distance. In that instance, mutual inductance is often unintended and harmful. Unwanted mutual inductance may also take place between adjacent traces on a circuit board or nearby wiring inside an equipment enclosure, this effect becoming much more pronounced at higher frequencies. This is one of the reasons that the wiring layout in a pictorial diagram differs radically from the schematic. In doing a repair on equipment that operates at a high frequency, it is important not to alter the routing of the wires. If it seems that they are not laid out in the most efficient way, this may be intentional to avoid parasitic inductive or capacitive effects.

In equipment that operates at high frequencies, it is important to avoid shortening or lengthening a wire, which could cause a characteristic impedance mismatch, resulting in harmful electronic reflections and data loss. (More about characteristic impedance in Chap. 10.)

Self-inductance, as the name implies, involves a single conductor with current passing through it. A magnetic field is established, and if the current is ac, pulsating dc, or some disorganized electrical activity that we may characterize as noise, there will always be a magnetic field surrounding the conductor. This field induces another current flow in the conductor, called "back emf." It is of opposite polarity at any given instant, and tends to oppose the original current. The bottom line is that this situation means that the conductor will exhibit impedance, specifically inductive reactance.

Inductive reactance (like capacitive reactance) together with plain old resistance, make up impedance. All of these are measured in ohms and all comply with Ohm's law. The two forms of reactance are frequency dependent, whereas resistance is not. Capacitive reactance decreases as the frequency becomes higher. Inductive reactance increases as the frequency becomes higher, and the formulas that are involved are similar but opposite in effect.

Multiple Circuit Elements

Resistance, inductance, and capacitance may coexist in a circuit, for example when one or more resistors, capacitors, or coils are placed in series or parallel, or any combination of the two. It is also possible, in fact inevitable, for any two or all three of these to be present in the same device. A resistor, for example, has a lead at either end in addition to any center taps, and these leads may function as the plates of a capacitor, the resistive material being the dielectric medium. Similarly, since the resistor is also a conductor, when there is current flowing through it there is actually a magnetic field around it, so that it is an inductance. These properties are very slight, but at high frequencies, the effects become significant to such an extent that the performance of the device, the circuit, and the entire piece of equipment may be compromised if not rendered completely nonfunctional if the design does not consider this effect.

The equation for inductive reactance is:

$$X_L = 2\pi f L$$

where X_L = inductive reactance
f = frequency (in hertz)
L = inductance

Notice the similarity to the equation for capacitive reactance, and the difference between them.

It seems to me that the universe may be characterized in a fundamental way as an intricate dance of particles, and their inscrutable interactions with elemental forces. Nowhere is this phenomenon more wondrous, in my view, than in the formulas that describe (define?) capacitive and inductive reactance.

The formulas are well established to the point of certainty, and as the decades roll on, we learn more about the fundamental processes, yet never get close to knowing why the particles and associated forces exist, as opposed to not existing.

Mutual inductance is similar, except that there is a second conductor within the area where the magnetic field is relatively strong.

If the magnetic field is fluctuating, current is induced to flow in the otherwise nonenergized second conductor. In this situation, the circuitry comprises a transformer.

The schematic in Fig. 1-9 perfectly portrays the electrical and magnetic properties of this ubiquitous device. There are no

FIGURE 1-9   Transformer.   mechanical moving parts except for associated items such as

cooling fans or pumps, or a circuit breaker or switch that could be built inside the same enclosure. What is moving is the magnetic field that surrounds either conductors or coils, and this is what makes the transformer work.

The two conductors are generally formed as coils wound around the same core. If this core is made of a material that is highly permeable, the effect of the mutual inductance becomes more pronounced.

An important and very common transformer application is for stepping up or stepping down voltage. The two windings in a transformer are the primary and secondary windings. The primary is the winding to which ac from an external source is applied, the input, and the secondary is where the induced voltage is available as an output. In theory, you could reverse input and output connections to convert a step-down to a step-up transformer, but such misuse would void the UL listing and could cause an unforeseen fire hazard.

Transformers are widely used in all sorts of electrical equipment. Isolation transformers make use of the fact that grounding does not pass through a transformer. If one conductor of the primary is grounded, none of the secondary conductors become grounded unless the windings are electrically connected. An isolation transformer is used in the power supply for a hospital operating room (with a line isolation monitor) to protect a patient who is undergoing invasive surgery from small ground currents that could cause injury. The primary and secondary windings generally have the same number of turns so that the voltage is neither stepped up nor stepped down.

The ratio of primary to secondary voltage is the same as the ratio of the number of turns in these windings. If the secondary has twice as many turns as the primary, the output voltage will be double the input voltage. This does not violate the law of conservation of energy. As voltage increases, the current decreases. Since volts times amperes equals watts (or volt-amps, for a reactive load) the power remains constant except for a small amount of energy lost in the form of heat, which is measured by a factor known as efficiency, expressed as a percentage.

The high-voltage winding may be identified by the fact that the wire, if accessible for inspection, is smaller. The larger wire is needed for ampacity to carry the greater amount of current. (The primary and secondary terminations should be clearly marked or self-evident.)

To understand the operation of a transformer, it must be realized that when a larger load is connected to the secondary, the primary will draw a correspondingly greater amount of current. In large equipment, these quantities may be conveniently measured with a clamp-on ammeter.

Transformers are generally reliable, but overheating or a manufacturing defect may set the stage for failure. Part of the troubleshooting process will entail looking at the transformer, especially when a power supply malfunction is indicated. There are other types of transformers in electrical equipment. These may be for coupling stages or impedance matching, and are often mounted directly on a high-quality loudspeaker. They are easy to test and, where defective, to replace.

High ohm readings indicate an open winding. In addition, when taken out of circuit, the windings should be found to be isolated from the metal core and enclosure, and from each other, unless it is an autotransformer (see following text). Beyond that, dc ohm readings are not definitive. It is possible that some turns within a winding could be shorted, changing the output voltage or overloading the primary. Either of these conditions could damage other components. The way to find out what is going on is to take in-circuit voltage readings with power to the primary. When performing this operation, it is essential to avoid danger of shock. This involves isolating yourself from dangerous voltage levels, especially when a step-up transformer is involved or when the equipment operates at 240 volts or higher.

One method is to clip your meter leads to the transformer terminals while the equipment is not energized, then power it up and observe the display from a safe distance. Remember to discharge all capacitors before hooking up.

Looking at Transformers

If the transformer is found to be defective, you have to ask why and what are the implications. One possibility could be an external power surge. This could burn out the primary or overpower the secondary, perhaps taking out downstream components in the process.

Another scenario involves a shorted component in the output circuit, which could place an excessive load on the secondary and, by induction, on the primary.

Electrical and electronic equipment frequently have power supplies, or low-voltage motor control or sensing circuits, and these generally require transformers, which should be checked out if diagnostic procedures point in that direction. A replacement transformer must be an exact match, in terms of input and output voltages and frequency, as well as power rating. Identical physical dimensions and construction details also have to coincide so that heat is not introduced in the wrong place to damage nearby components and so that there is not an impedance mismatch.

As for large power transformers, proper preventive maintenance will prevent outages and save future material and labor costs, as well as reduce fire hazard. Routine temperature readings, entered into a log so that damaging trends can be spotted early, should be part of the preventive maintenance program. Dirt or debris should not be allowed to accumulate to impede air circulation and cooling. Individual large transformer enclosures should be vacuumed out periodically, with power disconnected and locked out, and with measurements taken to ensure there is no backfeed. No one should undertake this sort of work without thorough training and certification in the safety aspects involved.

What Is an Autotransformer?

An autotransformer, as depicted in Fig. 1-10, has a single winding that serves as both primary and secondary. The autotransformer is smaller and less expensive than the dual-winding version, but electrical isolation is not provided.

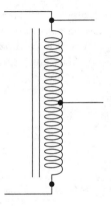

Figure 1-10
Autotransformer.

The most common application of an autotransformer is providing 120 volts from a 240-volt supply. When troubleshooting equipment where an autotransformer is part of the power supply, ohm readings will reflect the fact that primary and secondary windings are not electrically isolated. Grounding of one side of the primary will be conveyed to the secondary circuit, which is not the case in a dual-winding transformer.

There are safety implications in all of this. For one thing, insulation failure can cause the output circuit of a step-down autotransformer to exhibit full primary input voltage. Moreover, if the common part of the winding becomes open at any point (due to an unintended break in the wire), full input voltage will show up at the secondary.

Autotransformers are used in high-voltage power distribution systems to permit operation of machinery when the required voltage is not available. Using an autotransformer, you could run a 480-volt motor off an existing 600-volt source, and this would be a less expensive solution than employing a dual-winding transformer.

Another common application is in audio systems where it is desired to power speakers from a constant-voltage source, and to match impedances for a low-impedance microphone connected to a high-impedance amplifier.

The diode is an incredibly useful, widespread, yet simple device that is found throughout the world of electrical equipment.

Visually, it takes many forms, but may be identified by the fact that there are two leads, one often marked with a "+" sign, but otherwise having the appearance of a small resistor. In larger sizes, there is a prominent heat sink or even cooling fins. When forward biased, a significant amount of heat appears, and this energy must be conveyed away from the device so that excessive temperature rise does not occur.

The first diodes, other than cat's whiskers found in crystal sets, were power-hungry vacuum tubes, but in our solid-state epoch, the device is much simpler, more efficient, trouble-free, and less expensive.

A single diode will operate in one of two modes. Usually, with exceptions noted below, if it is forward biased, it will conduct. If it is reverse biased, it will not conduct. Biasing refers to the polarity of the voltage that is applied to the diode. Looking at the schematic, we see that of the two leads, the anode is connected to an arrow pointing from the lead into the device (Fig. 1-11) The other lead, the cathode, is connected to the other side, represented by a line segment perpendicular to the flow (or nonflow) of current. These terms, anode and cathode, are taken from the old vacuum-tube diodes. The bottom line is that when the anode is connected to the positive pole of a dc power source of appropriate voltage and the cathode is connected to the negative pole, the device will conduct, rather like a switch that is in the ON position. Then, the diode is said to be forward biased. If the polarity is reversed so that the anode is connected to the negative side, the diode is reverse biased and will not conduct. Accordingly, the diode may be considered a one-way gate, resembling a check valve in a water system.

You can build a very simple circuit to demonstrate the operation of a diode, as shown in Fig. 1-12 and Fig. 1-13.

In place of the pilot light, an ammeter may be deployed. If an ammeter is used, since it is a very low-impedance device, it is necessary to have a resistor or other load in series lest the diode be fried. Depending upon the relative polarities of the diode and dc source, the ammeter will indicate either the presence or absence of current through the circuit.

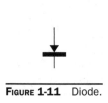

FIGURE 1-11   Diode.

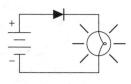

FIGURE 1-12   Forward-biased diode.

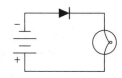

FIGURE 1-13   Reverse-biased diode.

To make the circuit work, diode, battery, and resistor ratings must be coordinated to avoid destroying the diode. Relevant information is in the manufacturer's data sheet, which can be downloaded off the Internet or obtained from the electronics parts distributor. As an exercise, you should acquire a data sheet for any active device that crosses your path, as these are easy to read and provide insight into how various devices work in circuits.

Of course, a diode is a sealed unit impossible to be repaired. However, it is instructive to understand its inner workings because such knowledge is useful in understanding and troubleshooting circuits containing one or more diodes.

Most semiconductors are made from silicon, which in its pure state is not much of a conductor.

An n-type silicon results from the application of a minute amount of phosphorous gas in a process known as doping. An atom of silicon has four electrons in its outer shell so that it bonds with four adjacent silicon atoms making a stable crystal. The outer shells of two adjacent silicon atoms in undoped state have eight electrons. This situation changes when a trace amount of phosphorous is added. Phosphorous has five electrons in its outer shell, and the fifth electron gains mobility so that it is free to move within the confines of the silicon crystal when voltage is applied. Since electrons are negative, the silicon that has been doped with phosphorous gas is known as an n-type material.

A cracked circuit board can be repaired. Straighten the board and reinforce it by gluing strips of plastic as needed. Solder light copper jumpers across any cracked traces. Be sure to use heat sinks on sensitive electronic components.

P-type silicon, on the other hand, is doped with boron, again in a gaseous state. Because boron has three electrons in its outer shell, it can bond with only three of the adjacent silicon atoms. Accordingly, one silicon atom has an empty space in its outer shell. This empty space is aptly called a hole. Unbelievably, these holes are charge carriers as well. The silicon that has been doped with boron gas is known as a p-type material because the hole is, in effect, a positively charged particle.

Inside a Diode

A diode is manufactured by fusing thin wafers of n- and p-type silicon to one another. The disc or cylinder so formed is equipped with leads at the two ends. Where the n-type and p-type material join is known as the junction. What is important in understanding any diode (or other semiconductor including the many types of transistors and integrated circuits) is the junction, and we have to consider what happens at the junction and on either side of it.

If the p-type side (the anode) is connected to the positive terminal of the dc power source, the charge carriers—holes—are repelled and pushed toward the junction. In this configuration, the negative pole of the dc power source is connected to the n-type side, the cathode, so the charge carriers are repelled and are pushed toward the junction. Then there is an abundance of charge carriers on both sides of the junction, and conditions are right for current to flow. The diode is said to be forward biased.

Conversely, if the connections are changed so that the positive pole of the dc power source is connected to the cathode (n-type material) and the negative pole of the dc power

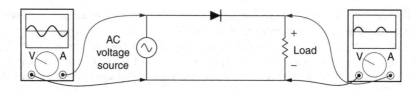

FIGURE 1-14   Half-wave rectifier.

source is connected to the anode, the two types of charge carriers—holes and electrons—are attracted to the dc power source poles. Then, there is an absence of charge carriers in the region of the junction. The diode is said to be reverse biased and it will not conduct, with an exception noted below. The junction becomes what is known as a depletion region.

As mentioned, there are two exceptions to the general explanation of the diode presented above. First, the diode does not immediately start to conduct upon being forward biased. The junction must see a certain minimum voltage before entering the conduction mode. This quantity is called the forward breakover voltage. It is generally less than 1 volt.

The other exception is known as the avalanche effect. What this interesting bit of terminology means is that in reverse bias mode, if the voltage rises above a certain level, the electrons will barge, so to speak, through the depletion region, and the diode will conduct reverse bias notwithstanding. This avalanche effect may be unintended and destructive, or it may be purposely incorporated in the circuit design, as in the case of the Zener diode, which we shall consider soon. Forward breakover voltage and avalanche voltage amounts are found in manufacturers' data sheets.

You should be aware that a diode might also act as a capacitor, with the junction becoming the dielectric, particularly at high frequencies. This effect may be unintended and harmful, or it may be part of the design.

That is how a diode works. There are variations depending upon how the diode is configured, packaged, and biased. Some of the applications are as follows.

Since the diode conducts only when forward-biased, that part of an ac waveform either above or below the x-axis will be eliminated, depending upon which way the anode faces in the circuit. This operating mode is enormously useful and easy to implement. Just connect the diode to either leg of the input in series with the load and you have built a half-wave rectifier (see Figs. 1-14 and 1-15).

The output is pulsating dc, with no electrical energy 50 percent of the time. This output would make quite the ac hum in audio equipment, but it can be put to good use charging a battery. A half-wave rectifier, while cheap and easy to build, has the disadvantage that it uses only half the power that is present in the ac circuit because it conducts during half the cycle.

FIGURE 1-15
Half-wave rectifier
output.

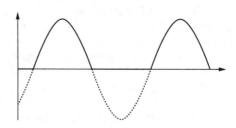

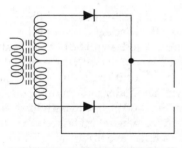

FIGURE 1-16 Full-wave rectifier.

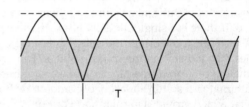

FIGURE 1-17 Full-wave rectifier output.

It can be used as a light dimmer, an improvement over the old rheostat-based device because excess energy is not dissipated as heat. There is just the heat loss involved in the forward-biased diode.

Another disadvantage is that the output of a half-wave rectifier is difficult to filter effectively due to high harmonic content. Furthermore, it is demanding of the diode because all the heat is concentrated in one place. In addition, it is harder on the transformer. For these reasons, a full-wave diode rectifier is often chosen (see Figs. 1-16 and 1-17).

Two diodes, with anodes connected to the legs of a transformer, are configured in parallel and the cathodes are brought together to make a positive dc output. The negative output comes from the transformer center tap. Most equipment with a dc power supply operates at less than line voltage, so there will be a transformer. Since the full-wave rectifier requires a center-tapped secondary and two diodes, it is a little more costly to implement, although the filtering is simpler.

Another variation, a full-wave bridge circuit, is shown in Fig. 1-18.

Notice that the cathodes of two of the diodes are connected to the + output and the anodes of the other two are connected to the − output. A center-tapped transformer is not necessary. Because there are four diodes, the heat dissipation is less problematic. The bridge circuit is also an improvement over rectifiers that are more primitive because it makes more efficient use of the transformer.

When troubleshooting, if in your opinion a power supply problem is indicated, look to the diodes. Where there is large current flow, consider excess heat, temperature rise, and resulting individual component failure as a cause of the equipment malfunction.

FIGURE 1-18
Full-wave bridge
rectifier.

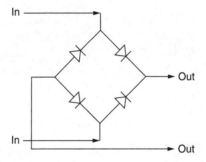

Lacking the schematic, you can tell a lot about a power supply by looking at the components. The power supply will be located near where the conductors enter the enclosure. If a significant amount of power is consumed, any diodes will be heavily heat-sinked. If there is a single diode, it is a half-wave rectifier. Two diodes with three secondary leads from the transformer indicate a full-wave center-tapped rectifier circuit. Three or six diodes may indicate a three-phase source. (Did you know that modern automotive alternators produce three-phase ac of varying frequency? Six diodes, one in each line inside the alternator, result in a reasonably smooth dc further refined when it is connected to the battery, which provides regulation.) Four diodes indicate a full-wave bridge rectifier.

In troubleshooting electrical equipment, power supply problems are frequently found, and diode replacement is often the answer. Besides rectification, there are other applications, but these are less frequent offenders because the power levels are less intense. Damage is still a possibility, however, because overvoltage may be applied for a variety of reasons. Applications are inside and outside the power supply proper.

Zener diodes take advantage of the avalanche effect. The avalanche voltage lies in the normal operating range so that when reverse biased above a certain value, say 60 volts, they will conduct without damage. This property makes the Zener suitable to function as a voltage regulator (see Fig. 1-19).

Thyrectors protect valuable electronic equipment, such as desktop computers and peripherals, from sudden, brief powerline transients or spikes (see Fig. 1-20).

You will notice that the thyrector consists of two diodes in series with polarities reversed. They are constructed as a single unit and connected between ac lines and ground. These devices are incorporated within a short plug-in strip with a resettable circuit breaker and power light, and sold as surge protectors. If the overvoltage is greater than a certain magnitude or duration, the thyrectors will be destroyed, sometimes making the electronic equipment go down as well.

Thyrector failure is seen frequently in areas of high lightning activity. Visual inspection and ohmmeter tests will often determine the status of the device. If there is any doubt about the condition of the device, replacement is recommended.

Some additional diode applications include:

- Electronic switching
- Envelope detection
- Step recovery
- Frequency multiplication

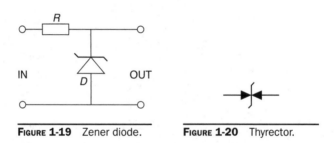

Figure 1-19   Zener diode.   **Figure 1-20**   Thyrector.

- Hot carrier devices
- Signal mixing
- Amplitude limiting
- Laser diodes
- Photo diodes
- Photovoltaic cells
- Optical isolators
- High-frequency oscillators (Gunn diodes and Tunnel diodes)

Sometimes a defective diode can be identified by its burnt appearance, but more often meter tests are needed.

Testing a Diode

The simplest way to proceed is to use a standard multimeter. It will be noticed that in ohms measuring mode, the meter will read either open circuit or circuit continuity (high resistance or low resistance) depending upon which way the meter probes are connected to the diode. This is because in ohmmeter mode, your meter of necessity applies a dc voltage across the component that is being tested. This dc voltage suffices to forward bias or reverse bias the diode, a rather unscientific but adequate way to proceed. The actual ohms values do not mean anything.

Most but not all meters apply positive voltage to the red probe. You can determine the polarity by connecting a known good diode to the meter. When the positive pole is connected to the diode's cathode, as indicated by marking, the ohmmeter will read low resistance. Then, if necessary, you can color code your meter probes again.

You can make a dc polarity tester by soldering a diode into a spare probe lead. This accessory makes a useful tool for checking circuitry, particularly automotive, boat, and aircraft wiring.

Of course, the ohmmeter test is not fully definitive. For one thing, it does not check for forward breakover and avalanche voltage to see if they are within specifications, as shown in the data sheet. Furthermore, real-world operating parameters at elevated voltage, current, and frequency conditions may not be acceptable.

Multimeters with advanced features sometimes include a "diode-check" function, whereby a low-level current is passed through the diode so that voltage drop can be displayed.

Fortunately, most low-power diodes are cheap enough so that you can afford to install a new one to see if that is the problem. This does not address the problem of corollary damage to some other component accompanying diode failure.

This completes our discussion of diodes for now. The next topic builds upon the foregoing subject matter. More advanced semiconductors, including transistors and integrated circuits (ICs), are semiconductors like diodes but operate in a somewhat more complex manner. Instead of one p-type and one n-type material meeting at a single junction, there are three or more layers having various characteristics and multiple junctions. Using these concepts, researchers (especially in the mid-1900s) developed a multitude of different active components, which made possible advanced types of audio and video equipment,

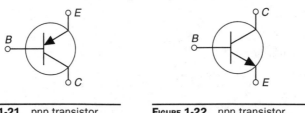

FIGURE 1-21   pnp transistor.   **FIGURE 1-22**   npn transistor.

computers, telecom networks, and much more. These active components can be diagnosed and replaced once the basic principles are understood. The place to start is with an understanding of solid-state electronics. This may be acquired through practical experience in the field, in books that are readily available, and, of course, on the Internet.

Like diodes, transistors are composed of p-type and n-type material. A bipolar transistor has three layers with two junctions. The pnp transistor is shown in Fig. 1-21 and the npn transistor is shown in Fig. 1-22.

The center region for both types of bipolar transistor is called the base. The line having the arrow is always the emitter, so in a schematic diagram it is not necessary to label the terminals. If the arrow in the emitter points in toward the base, it is a pnp transistor. If the arrow points out away from the base, it is an npn transistor.

In most applications, npn and pnp transistors work the same, with one major difference—all polarities are reversed and the direction of current flows are opposite, both inside and outside the device. As long as the transistor specifications (as given in the manufacturer's data sheet) are equivalent, pnp and npn transistors are usually interchangeable provided voltage polarities are reversed.

Transistors Have Multiple Uses

Transistors have many applications. The most basic and commonplace is amplification, as in an audio amplifier. An example of amplification in everyday life is when you are driving a car and slight changes in the position of the gas pedal due to foot pressure are instantly translated into variations in engine rpm and vehicle speed. In fact, motion of the gas pedal is the result of a still slighter level of electrical activity in your brain and nervous system. The totality is a type of multistage amplifier, where the output of the first stage is coupled to the input of the second stage. Unlike the step-up transformer that we considered earlier, where there is no increase in power (just a rise in voltage with decrease in amperage), in amplification the power is actually increased. The power is multiplied many times from the input of the first stage to the output of the final stage. Here again, the law of conservation of energy is not contradicted. At each stage, power is fed into the system from an external power supply.

In the automotive example, it is the gasoline engine that powers the car. The motion of the gas pedal controls the output level, but the actual energy comes from elsewhere.

The three terminals of a bipolar transistor allow for a two-wire input and two-wire output, with one terminal common to both circuits.

The varying input provides changing bias on one junction, and a larger dc power source in series with the transistor output provides bias for the other junction. When both junctions are forward biased at the same time, the transistor conducts at a level greater than the input.

In common emitter configuration, the control circuit terminates at the base and emitter. Here a small current controls the larger current that flows in the collector-emitter circuit. For both pnp and npn transistors, the flow of electrons is in the direction opposite the arrow. The direction and placement of the arrow tell you the type of transistor and identify the terminations.

The common emitter configuration is fundamental and easy to understand. There are common collector and common base applications as well. From the schematics or visual inspection of the equipment, you can ascertain which lead is common to the input and output circuits, but this approach is limited by the fact that there is no universal standard for marking, color-coding, or positioning the three leads of a bipolar transistor. You would think that the middle lead would go to the base, but this is not always the case. Fortunately, there are tests that can be performed with an ohmmeter. There are full-featured transistor testers as well. These fall into three categories.

A quick-check in-circuit transistor tester measures the transistor's ability to amplify a signal. Without removing the transistor from the circuit, a rough measurement is made so that you can decide if replacement is warranted.

A more sophisticated instrument is the service-type transistor tester. It ascertains the forward current gain (known as beta) of the transistor. It also checks for base to collector leakage current with no current going into the emitter. Additionally, some service-type transistor testers identify the leads—collector, emitter, and base.

The high-end transistor tester is known as a laboratory-standard transistor analyzer. It will simulate a real operating environment, providing voltage, current, and signal inputs and looking at the output.

These testers come with extensive operating instructions, and you will find all of them user-friendly. Let us assume for now that all you have is a simple multimeter with ohms function. It will be possible to check a transistor, although the amount of information provided regarding its actual status under dynamic operating conditions would be less than total. However, it is a good way to proceed.

Transistors are incredibly reliable and rarely go bad. Many transistors have been in service for over 50 years, and they are still as good as the day they were soldered in place. Nevertheless, it is possible for a transistor to fail if excessive voltage or excessive heat is applied to it. Static electricity, line surges, high ambient temperature, failure of a ventilating fan, inadequate heat sinking, or external circuit conditions may cause transistor failure. Simple ohmmeter checks will let you know the status of the device on a go, no-go basis.

As mentioned in the discussion of diodes, a multimeter in the ohms mode is useful for testing a solid-state device because the dc power supply serves to bias the NP junction. Naturally, the full power supply voltage is not applied. The ohms value as shown in the readout is not to be taken as a meaningful resistance measurement, but it does indicate whether the holes and electrons have migrated to the junction so that the particular junction can conduct, given the amount of bias provided by the meter.

The first task is to determine the polarity of your meter's probes when in the ohmmeter mode. Most meters are configured so that the red probe, when connected to the ohmmeter function socket, is positive, and the black probe, connected to the common socket, is negative. However, meter manufacturers have not been consistent, so you have to test the tester. The procedure is to find a known good diode with polarity marked on it. The diode could have a diode schematic symbol printed on it, a single band on the cathode, or + and – signs. Orient the diode so that it is conducting according to your meter. The probe that is connected to the cathode will be positive because it is repelling the holes and pushing them into the region of the junction. Then, permanently identify the polarity of your meter so that you can make meaningful diode and transistor tests and distinguish between npn and pnp transistors, and identify the leads if they are not marked. We have outlined the procedure for testing a diode. Testing a transistor is more complex. Keep in mind these points:

- Solid-state component manufacturers package, terminate, and mark their products in a variety of ways, so make no assumptions. A bipolar transistor will have three leads, but it may have no identifying color code or markings. It is even possible that the middle lead will not be connected to the base, although that would be logical. Assuming that you do not have an overall schematic of the piece of equipment or data sheet for the component, you have to start from scratch. (If you do have the schematic, you will know that the terminal with the arrow is the emitter and if the arrow points inward, then the component is a pnp device.) By the circuit wiring or printed circuit board traces, you should be able to identify connections and biasing of the other terminals.

- If the component is in circuit, adjacent components can cause misleading meter readings, so it will be necessary to temporarily cut or desolder some terminations. You can always leave one lead attached as it will not comprise a complete circuit path and your readings will not be compromised. When desoldering or resoldering these connections, remember that heat is the enemy of solid-state components. Use a minimum of heat to accomplish the operation and protect the component by using a clip-on heat sink. (More about this later in this chapter when we talk about repair methods and materials).

- When working with solid-state components, you must protect them from static electricity. Your body can acquire a static charge from time to time. This can be controlled. A copper grounding bracelet connected to the system ground will take care of this charge. Similarly, your soldering iron or any other metallic tool could introduce an unacceptable voltage. Frequently touch it to a known ground.

To test a bipolar transistor using a multimeter, now that you have ascertained the polarity of the ohmmeter probes, proceed as follows.

A transistor may be thought of as two diodes, back to back. If the transistor is npn, the two anodes are connected and the base connection is tapped from where they join. If the transistor is pnp, the cathodes are connected and the base is tapped from where they join (see Figs. 1-23 and 1-24).

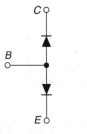

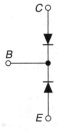

FIGURE 1-23 Diode equivalent of npn transistor.

FIGURE 1-24 Diode equivalent of pnp transistor.

It must be emphasized that these diode hookups will not function as working transistors. They are just models that indicate the meter connections for testing transistors.

It is possible that you do not know whether the transistor is npn or pnp, and it is also possible that you do not know the identity of the three leads if you do not have the schematic or the manufacturer's data sheet. You can nevertheless perform tests with an ohmmeter, taking advantage of its ability to simultaneously bias and measure continuity of each pair of leads.

Since there are three leads, there are three pairs that may be tested, and since each pair can be biased in either of two directions, it means that there are six possible readings that will provide the only information to be acquired using your ohmmeter.

It is possible to draw some conclusions. The collector and emitter should read open regardless of which way they are biased, because the two conceptual diodes are pointing in opposite directions. As long as the avalanche voltage is not reached, they will not conduct.

If there is low resistance on all three pairs, regardless of biasing, the transistor is shorted and defective. If there is high resistance on all three pairs, regardless of biasing, the transistor is open and defective.

Assuming that the transistor thus far tests well, you have identified the collector-emitter pair, so you know the remaining lead is connected to the base. The base will conduct to either of the other leads when forward biased, and will not conduct when reverse biased. For these two readings, the biasing is opposite. In other words, to make an emitter-base junction conduct, the anode must be connected to the base if it is a pnp device. If you know whether it is an npn or a pnp transistor, you can determine which lead is the emitter and which one is the collector. If you know the identity of one of these leads, then you will know the identity of the other because the base has already been identified. From this, you can deduce whether the device is npn or pnp. (All of this assumes that you know the polarity of your ohmmeter.) It is not possible with these simple tests to identify the leads, aside from the base, if the transistor type is not known, nor is it possible to determine the transistor type if you do not know which lead is the collector or which is the emitter.

Tests that are more sophisticated will reveal both of these unknowns, taking advantage of the fact that the emitter material is more heavily doped and so will have a lesser voltage drop.

With the tests outlined above, however, we have obtained enough information to decide quite often whether the transistor is defective, if not all of its parameters. These tests are preliminary and are not to be construed as providing a definitive dynamic analysis. However, transistors usually fail all at once if at all; they do not gradually weaken.

Using the Diode Check Function

Some meters incorporate a diode check function, and this displays voltage drop rather than the pseudo-resistance reading you get from an ohmmeter. Therefore, where possible, you will want to use the diode check function.

As you might expect, a transistor will amplify the input voltage only up to a certain point. After that, the output levels off and the transistor is said to have reached its saturation point. In analog applications, where it is desired to amplify a signal without distorting it, the transistor is kept below its saturation point. In contrast, digital circuits usually involve transistors that are biased beyond saturation when in the high state, and biased so as not to conduct when in the low state. These two states represent the binary numbers 1 and 0.

Two very important transistor parameters are alpha and beta. In a bipolar transistor, alpha is the collector current divided by the emitter current with base at signal ground. This quantity is the dynamic gain of the transistor. In contrast, beta is defined as collector current divided by base current when the emitter is at signal ground. Both of these parameters assume a small electrical signal applied at the input. Two formulas that illustrate the relationship between alpha and beta of a given bipolar transistor are:

$$\text{Alpha} = \text{Beta}/(1 + \text{Beta})$$

and

$$\text{Beta} = \text{Alpha}/(1 - \text{Alpha})$$

The underlying current relationship is:

$$I_C = I_E = I_B$$

where I_C = collector current
I_E = emitter current
I_B = base current

Bipolar transistors may be connected in any of these configurations: common emitter, common base, and common collector. Each of these can be designed to operate as an amplifier.

The common emitter configuration, when used as an amplifier, provides high gain. It must be remembered that the output is always 180 degrees out of phase with the input, but this is not a problem as long as we know what to expect. (An even number of successive stages will result in a noninverted output.) Aside from the phase change, a common-emitter amplifier provides excellent fidelity when operated below saturation.

The common-base configuration is similar, however, with base at signal ground. Output current appears at the collector so that amplification occurs. The output is in phase with the input, not 180 degrees out of phase as in the common emitter configuration.

The common-collector hookup, also called the emitter-follower circuit, involves placing the collector at signal ground. There is no amplification. The circuit is used to provide isolation between other stages, and for this reason, it is known as a buffer. Input and output signals are in phase. The only configuration where output is out of phase with respect to input is the common-emitter circuit.

Amplifiers are divided into four classes, depending upon how they are operated: Class A, Class AB, Class B, and Class C.

Class A amplifiers are biased so that they are never operated below cutoff or above saturation. Thus, collector current flows during the entire input. Class A operation is appropriate for many applications including audio and RF amplification.

Class AB operation is characterized by the fact that the forward bias voltage is less than the input signal peak voltage. The base-emitter junction is reverse biased for part of each cycle. Collector current, accordingly, flows between 180 degrees and 360 degrees of the input signal. Class AB operation is employed in push-pull amplifiers because it mitigates the harmful effect seen in Class B operation known as crossover distortion.

Class B operation is biased so that the collector current is cut off for one-half of the input cycle. Its application is for audio amplifier final stages, where the output power is very high, and for transmitter power-amplifier stages.

Class C operation occurs when collector current flows for less than one-half of the input cycle. It is employed for transmitter radio-frequency amplification.

Thus far, we have been talking as though the bipolar transistor were the only type. This is not at all the case. In fact, there is a great variety of transistors and other semiconductor devices that you will see in all types of electrical equipment. They operate on the same fundamental principles as the diode and bipolar transistor, but typically the structure is a little more complex or, in the case of ICs, vastly more so. There are often four or more terminals, making for more complex circuitry inside and outside of the device, with enhanced functionality. We shall look at a few of the varieties frequently encountered in troubleshooting and repairing electrical equipment.

Other than the bipolar device that we have been discussing, another widely used type of active device is the field-effect transistor (FET). Two types are the junction FET (JFET) and the metal-oxide-semiconductor FET (MOSFET).

The JFET, like the diode and bipolar transistor, is composed of n-type material and p-type material. The parts of the JFET have highly descriptive names—gate, channel, and drain. There are two varieties of this device—the n-channel and the p-channel.

Electrons or holes, as the case may be, move along the channel between the source and drain. This comprises a flow of current that is of the same magnitude at the source, at the drain, or at any point along the channel. A fluctuating electric field within the solid-state material causes the channel current to fluctuate. Small fluctuations in the gate voltage are what cause the field that envelopes the channel to also fluctuate. The bottom line is that there are corresponding changes in the channel current that occur at a much higher level due to external biasing. Actually, the higher the gate voltage rises, the lower the channel current falls, so it is an inverse relationship.

To summarize, the voltage at the gate produces an electric field that chokes off and limits the channel current—electrons for an n-channel JFET and holes for a p-channel JFET. Looking at Figs. 1-25 and 1-26, you can tell whether it is an n-channel device, with the gate arrow pointing inward, or a p-channel device, with the arrow pointing outward.

In some schematics, the arrow is missing, but you can discover the JFET type by looking at the power supply polarity.

A depletion region forms in the channel in response to any increase in gate voltage. For an n-channel JFET, this will be in a negative direction and for a p-channel JFET, it will be in a positive direction. Either way, increased voltage at the gate means decreased current along the channel and at the output of the device. This is because a depletion region forms in the

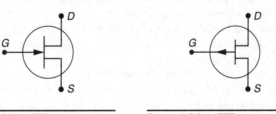

FIGURE 1-25   JFET, n-channel.   **FIGURE 1-26**   JFET, p-channel.

channel as a result of the more intense electric field. When the field intensity reaches a certain level, the channel current falls to zero. This event is known as pinchoff.

All of this sounds like the performance of a bipolar transistor, with the difference being primarily in terminology. However, there are a few substantial differences. FETs are unipolar rather than bipolar. The channel is a homogenous substance without a junction, and the current flows from one end to the other. Looking at Figs. 1-25 and 1-26, you can see that these two terminals—drain and source—appear the same. In fact, in many applications, it is possible to connect them either way and the device will perform in the same manner.

A major difference is that whereas in a bipolar transistor the input is a fluctuating current, in the JFET the input is a fluctuating voltage. The amount of current required to establish and fluctuate the electric field in the very small channel is minute. Therefore, we may say that the input of the device is very high impedance. It loads the circuit that feeds it to a very small extent, and is practically invisible to it. That property makes it a very useful device. For example, it may be used in a voltmeter to create a high-impedance instrument, so that the circuit being tested will not be loaded and altered because of the measurement being taken.

What Is a MOSFET?

Like the JFET, the MOSFET can be used either for switching or for amplifying signals. It further resembles the JFET in that it has a gate, source, and drain. Moreover, there can be either a p-channel or an n-channel. MOSFET, in a way, is obsolete terminology because metal oxide is no longer always used. Another term, perhaps more descriptive, is insulated gate field effect transistor, but most technicians still use the original acronym.

The main difference between a MOSFET and a JFET is that in the former the gate is not directly connected to the semiconductor material. Separating them is an extremely thin silicon dioxide insulator. Therefore, the connection is actually a capacitor, passing higher frequencies with less impedance and capable of holding the very small charge involved for some time.

There are two types of MOSFETs—enhancement mode and depletion mode. The enhancement mode is far more common. The difference between the two is all in the biasing. The enhancement mode is off when the voltage between gate and source is zero, and on when the gate voltage moves in the direction of the drain voltage. A depletion-mode MOSFET, on the other hand, is on when the voltage between gate and source is zero, and will be turned off by the application of voltage.

Like the JFET, the MOSFET has very high input impedance, practically infinite. While this is useful where you do not want to load the circuit connected to the input, it also means that the device is very sensitive to static charge, which can destroy it without changing its outward appearance. MOSFETs come packaged in conductive foam in which the terminals are imbedded, effectively shorting out the internal elements so that there can be no damaging static charge buildup. It is a good practice to leave the MOSFET in the conductive packaging up until the time you are ready to install it. Moreover, do not touch the leads if your body may be carrying a static charge.

In testing a MOSFET, you have to remember that if voltage is applied to the input, the MOSFET will be turned on or off (as the case may be), but if the voltage is removed, the MOSFET will not revert to the nonenergized state due to internal capacitance. To make it revert, you have to bleed off the charge using an appropriate resistance.

Most multimeters, in the diode check mode, will apply about 3 volts, which will not harm the device and yet will be sufficient to activate it.

To begin, connect the negative probe to the source. Then connect the positive probe to the gate. The next step is to shift the positive probe to the drain. You will read low because the meter has charged the gate capacitance. Leaving probes connected as above, short out the source and gate, using your finger. The gate will become discharged with respect to the source and the meter will read high, indicating that the device is not conducting. All of this activity indicates that the device is working.

The test does not reveal all the parameters given on the data sheet, but it will let you know if the device has failed.

Often a defective MOSFET will have a burnt appearance. When these devices short out between drain and gate, relatively heavy current may be fed to other MOSFETs and drive transistors, so it is necessary to check out everything.

Diodes and transistors are rugged and well protected within sealed packaging. However, external forces, including environmental heat and excessive current or voltage, may damage a solid-state component. It is usually cheap and easy to replace.

Many electricians and electronic technicians simply replace an entire circuit board, and this may be the best option if time is of the essence and money is no object. However, complete circuit boards are costly and there is the serious problem that if you go this route, essentially handing off the repair to another organization, you have less control. It is possible to invest in a new circuit board and replace the old one, only to find that the problem persists. For the cost of a few circuit boards, you could keep an inventory of a great many transistors, diodes, capacitors, and resistors. With the proper techniques, soldering new components in place is not difficult, with the exception of ICs. We will discuss actual repair procedures later.

Thus far, the focus has been on discrete components—devices packaged individually with leads or terminals attached to internal elements to facilitate external connections. There is another category known as the integrated circuit or microchip. It is widely used and, in fact, our highly wired world of today depends heavily upon it. ICs are everywhere, from a simple household appliance to the International Space Station.

This ubiquitous item is a very small plastic or metal solid with conductive pins attached, sometimes 14, sometimes more or less, so that it may communicate with the outside world. ICs are made up of many very small electronic components connected together to form an electrical network. The components, transistors, resistors, capacitors, and so on are placed on a monolithic semiconductor slab that forms one pole of some of the components, while others are electrically isolated from it by means of an insulating layer.

A process known as photolithography creates the whole thing. The circuitry is enormously complex, sometimes involving over 1 billion components. The design process, of necessity, is highly automated because no one human could conceive of complexity on this scale. An IC could never be opened up and repaired, although replacement is a routine task for electronic technicians. If you can isolate the fault first to a board, then to a circuit, and finally to a defective IC, a replacement should get the equipment up and running unless one or more other components were overloaded concurrently or preceded the demise of the IC.

There is a problem replacing ICs when they are soldered onto a circuit board, as opposed to being plugged into a socket. To take out the old IC, it is necessary to melt the solder joint for every pin simultaneously, in order to pull the component. The same is true for installing the new one. So much heat can damage the semiconductor. Fortunately, there are tools and procedures to facilitate the operation, and we will discuss them later.

There are limits to how an IC can be used. For one thing, inductors cannot be built inside the chip, so any coils have to be deployed outside the package, and wired appropriately. Second, ICs can be used only in very low-power applications because any heavy current within the component would produce excessive heat in such a small volume and the temperature rise would quickly destroy the IC. Otherwise, these chips are very useful and found in all types of equipment. Heat sinking and cooling by means of a fan or natural convection help control temperature rise.

ICs may be linear, for analog circuitry such as found in a stereo amplifier, or nonlinear, for data processing. The physical principles are the same, in terms of semiconductor behavior, but circuit details and biasing are different.

Some of the applications for ICs are:

- Voltage regulator, part of a power supply that ensures the voltage level will remain constant despite loading variations, within specified limits. To identify a voltage regulator IC, look for three terminals.

- Timer, often seen in conjunction with a potentiometer, which may be adjusted to control the delay. The timer consists of an oscillator with an electrical output at a specified frequency and means for counting the pulses so that the length of the delay may be set.

- Multiplexer, combining two or more signals so that they may be distributed on fewer lines. There are numerous multiplexing schemes, each associated with a demultiplexer. Telephone transmission upstream from the branch exchange involves this technology.

- Computers rely on random-access memory (RAM) and read-only memory (ROM). These are two distinct areas that coexist within most computers. A detailed knowledge of how they interrelate and work together is necessary in computer repair work, and we will look into some of the details in Chap. 8. ROM exists within dedicated ICs.

- Operational amplifiers work in a rather unique way, and because of their properties, they are quite plentiful in electronic equipment (see Fig. 1-27).

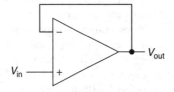

FIGURE 1-27
Operational amplifier.

A single output derives, at an amplified level, from two inputs, one inverting and the other noninverting. There are two power supply connections. One is connected to the emitters and the other to the collectors.

Outside the package, a resistance is connected between output and noninverting input, making for positive feedback and, at a high level, oscillation. Between the output and the inverting input, circuit elements are introduced with values chosen to cause amplification to vary with frequency. This occurs because the feedback line contains sufficient capacitive reactance to vary the feedback in accordance with the signal frequency.

The inverting and noninverting inputs are what make an operational amplifier function as it does, and these properties are well suited for treble-bass adjustment.

ICs may be tested, but the procedure is complex and requires specialized equipment. As a final stage in the manufacturing process, ICs are tested and certified. When equipment is first placed in service, it is assumed that all ICs will operate within specifications. Subsequently, excessive voltage or heat, whatever the source, may damage a sensitive component inside the package or indeed the entire semiconductor substrate. The damage may or may not be visually apparent. In the latter case, we must find ways to verify the operation of the IC in-circuit.

The best test of an IC is functional. That means observing it in circuits that will provide inputs and measuring the output voltages or looking at the waveform.

Most ICs cannot be effectively tested with a multimeter. There are highly specialized testers that are capable of testing a number of ICs, but this equipment is very expensive and you probably will not have access to it when you need it, unless you work in a shop or laboratory that does advanced work.

However, the good news is that many IC manufacturers have data sheets that include simple test circuits that you can build.

There is a test circuit for the very important 555 timer IC (see Fig. 1-28).

Of the many ICs available, the 555 is one of the simplest, yet it is very useful. A lot of electrical equipment has blinking LEDs that indicate operating states or a failure mode, and it is likely the 555 is at work. It was designed in 1971 and at present, there are variations from different manufacturers. Typically, the silicon substrate has 25 transistors, 2 diodes, and 15 resistors. The package is an 8-pin dual-in-line metal or plastic rectangle. There is no need to comprehend the internal circuitry, but to successfully troubleshoot electrical equipment containing a 555, you should know the operating modes, pin connections, and associated external circuitry.

There are three operating modes: monostable, astable, and bistable.

In the monostable mode, a single pulse is generated. One very useful application is the bounce-free switch, where if a switch such as a keypad element should inadvertently create two or more bogus pulses, the 555 makes it right.

In the astable mode, the 555 functions as an oscillator with a very wide range of intervals between pulses, determined by external circuit components. This is useful for making an LED blink.

In the bistable mode, the 555 becomes a flip-flop, with output in the form of rectangular pulses, the frequency determined by the external components.

FIGURE 1-28
Test circuit for
555 timer IC.

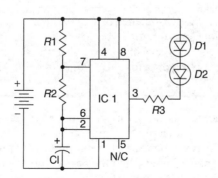

Varieties of Test Equipment

Testing of ICs is best done on a case-by-case basis. This means that you cannot have a single test circuit or meter that will work for all ICs. Some ICs are quite expensive, and it is not feasible to substitute them on a trial-and-error basis. Remember that heat or static electricity will destroy these chips, so proceed with caution.

We have discussed some of the electronic components that you will encounter and how to recognize their failure modes.

In doing this sort of work, test equipment is essential. Without it, one would be in darkness, as far as what is happening electrically. For quick measurements of incoming power, to determine presence or absence of voltage, and to get a general idea of the level, for example, 120 or 240 volts, the neon test light is a great little tool. It is inexpensive and clips onto your shirt pocket like a ballpoint pen, so you will have it when you need it. The principle is that two electrodes inside the small glass envelope ionize the inert gas, so that when voltage is applied, a faint yet distinct glow is emitted. It is a very high-impedance device. A small internal resistor further limits the current flow. If the probes are touched to a 120-volt hot leg and ground, the bulb will glow and the light is bright enough to be seen outside in full daylight. If the probes are touched to 240-volt terminals, the neon bulb will glow twice as brightly, so you can jump around among terminals and quickly find out what is what.

A common failure mode in large commercial appliances as well as in distribution boxes, meter sockets and entrance panels is loss of one phase, which will make equipment work erratically if at all. The neon tester is excellent for finding a dead leg.

You can touch one probe to a hot terminal and the other to your finger, and the bulb will indicate presence of voltage. This is an intrinsically safe operation because the neon bulb plus resistor has such high impedance that virtually all the voltage appears across the tester and minute current flows through the body. However, the electrician must always be aware of issues that may be involved, such as standing in water or introducing current through a cut or abrasive wound on the hand. (Dry, clean skin, especially of a worker with toughened hands, is a partial insulator.) In addition, if the tester has become shorted, you will be zapped, so you should touch the free probe to a known ground rather than to your finger. The technique just described is useful for determining which line is hot and which is neutral. In addition, before and after completing a repair on a tool or appliance with a metal case, you can check to make sure the enclosure has not become energized due to an internal fault.

A typical neon tester is rated for up to 600 volts, but that is a little too close for comfort.

> A pencil eraser will remove high resistance oxide coatings on electrical contacts in a motor controller, TV remote, or anything in between. Rub the eraser vigorously on clean paper until the shelf-life glaze is gone, and then polish the contacts. This will not remove severe pitting or burning but it works for early stages of oxidation and is great preventive maintenance.

Another convenient tool for checking for presence or absence of voltage is the humble appliance bulb in conjunction with a plug-in bulb socket. This device is good for checking receptacles to see if they are live, whereas voltmeter probes are problematic for this measurement because you never know when they are getting a good connection inside the receptacle. Plug-in bulbs can be left in one or more receptacles while you work elsewhere on connections some distance away, providing indication when the branch circuit goes live.

The next step up in order of increasing sophistication is the so-called Wiggie, also known as a solenoid voltage tester. It has several advantages, making it popular with electricians. Besides a visual voltage indicator, it makes a buzzing sound on ac and a loud click on dc, helpful when it is difficult to watch the indicator and place the probes simultaneously. Moreover, the buzz or click can be felt if you are holding the instrument—a good feature in noisy environments. The Wiggie does not require batteries for operation—a definite plus. Wiggie is a Klein Tool registered trademark, but other manufacturers such as Square D carry a similar model.

The circuit analyzer, made by several manufacturers, plugs into a receptacle to indicate presence or absence of voltage. Additionally, three LEDs indicate whether wiring is intact and connected correctly. Major wiring errors are displayed by the LEDs, with a key printed on the plastic case. These include open ground, open neutral, and reversed polarity.

The Ideal SureTest Circuit Analyzer is a much more full-featured instrument that sells for over $300. It is actually a U.S. version of the European loop impedance meter. It applies a load to a circuit under test and reveals various circuit parameters. The meter will provide a wealth of information about circuit status including wiring that is concealed behind walls such as improper grounds and intermittent faults such as occur when a nail has partially penetrated a wire causing circuit failure only under heavy load. It also reads out line, ground, ground to neutral, and peak voltage, as well as frequency.

Megger makes an excellent loop impedance tester that checks the integrity of the equipment-grounding loop and generates a report that can be downloaded to your computer via a USB port and printed for purposes of documentation and certification (see Fig. 1-29).

Figure 1-29
Megger's loop impedance meter. (*Courtesy of Megger.*)

The common multimeter is the electrician's most-used measuring tool by far. Older models were analog, with a needle pointing to the correct value when the relevant scale had been selected. The analog meter is more difficult to read and perhaps less rugged, but it is still preferred by some electricians as it may perform better outdoors in very cold weather.

The digital version, as the name implies, has a numeric readout that responds to input from the probes.

Some models have diode check, capacitor check, and other capabilities. The most basic model reads voltage, resistance, and low-level current. FETs on the front end of the voltage function ensure that the circuit under test will not be loaded to any appreciable degree, so that if it is a high-impedance source that is being tested, the voltage will not be dropped.

An inexpensive meter will test up to 600 volts, but great care must be exercised when measuring live circuits. There are several ranges and the idea is to start on a higher range and work down if there is any doubt as to the maximum possible voltage. Some meters are auto-ranging, so you do not have to worry about that. For dc voltage measurements, polarity must be observed, although some models will sense polarity and switch poles internally if necessary. If OL appears in the readout, it stands for overload and means that you have to select a higher range. However, sometimes OL will appear when the meter is not connected to a live circuit. How is this possible?

Many beginning electricians are puzzled by the phenomenon of phantom voltage. The fact is that there are static charges at times in every conductive body that is or has been in moving contact with a nonconductive body, including dry air. There is electrical potential everywhere, even in the meter case and probes, and a very high-impedance instrument such as a digital voltmeter may see a high voltage drop for no obvious reason. As soon as you connect to a live circuit, the phantom voltage reading will go away and your meter will lock on to the real quantity.

If your meter reads current, it will only be in milliamp ranges. This is because the current has to pass through the meter in order to be measured. The power that has to be dissipated is equal to I^2R, so with any large current, the meter could not handle it. For this reason, the current function is rarely used.

Do not forget that to take a current measurement you have to temporarily cut open the circuit in order to place your meter in series with the load. This is in contrast to a voltage measurement, which is taken in parallel with the load or with the source. In current mode, the meter has very low impedance and is invisible to the source and load. In voltage mode, the meter has very high impedance and is invisible to the source and load.

A very useful function, perhaps employed more frequently than any other, is the ohmmeter mode. There should be no mystery in how to use it. For ordinary resistance measurements, where the object of interest is not an active component that needs to be biased, polarity does not matter. However, be sure the test voltage injected into the component or circuit, usually about 3 volts, will not cause damage. It is good to have two meters so you can test this voltage on both meters.

Some meters have an audible beep or buzz indicating that the reading is fewer than 20 ohms. This is very convenient as a continuity tester. For 0 ohms, the readout may say OL, which in the ohms mode means Open Loop.

In the resistance mode, any voltage may damage the meter. Some instruments have an audible alarm or internal protection, sometimes in the form of a fuse that would have to be changed.

The only drawback for this instrument is that it depends upon an internal dc power source, often a 9-volt battery. With moderate daily use, the battery will last about a year. Immediately before measuring hazardous voltages to see if equipment is indeed deenergized and safe to work on, it is necessary to check the meter by measuring a known live voltage. The battery not only supplies test voltage for the ohms function, but also biases for internal electronics including the readout.

When running multiple parallel conduits across a ceiling or wall, make a plywood template based on the originating panel knockouts to aid in maintaining uniform spacing. Secure conduits to struts without fully tightening strut clamps at first, and then check alignment at each strut and tighten. Use the same template for drilling holes through an existing wall.

To do any serious troubleshooting on a commercial level, you will definitely need a clamp-on ammeter. Older models were analog and did not need a battery. These work well, but a digital model is now available and it is very popular with electricians because it includes voltage and resistance functions, and a hold button so that the readout will clamp on the highest current since the feature was activated.

Unlike a multimeter, the clamp-on ammeter will measure very heavy currents, as seen at an electrical service.

The clamp-on meter, rather than measuring current directly, measures the magnetic field surrounding a conductor. Just open the jaws, pass the conductor into the space between them, and close the jaws. The readout is surprisingly accurate, in either the old analog type or the new digital model. It does not matter if the conductor is centered in the space between the jaws, and if you move it around, the readout remains consistent.

It is necessary to realize that you cannot take a reading off a supply cord that contains two hot legs or a hot and neutral. This is because the current flows in opposite directions, so the magnetic fluxes surrounding the conductors cancel out, giving a zero reading. In this situation, you may be able to clamp onto an individual current path inside the equipment, or back at the entrance panel. Alternatively, you can make a splitter. Make up a short extension cord with a male plug and a female connector at the two ends. Slit the outer jacket for about 8 inches near the middle and pull out the wires. Trim back the cut insulation and you have a very useful accessory to the clamp-on ammeter. You can make noninvasive current checks of tools and appliances.

For motorized equipment, the current draw should be less than the full load current on the nameplate, except while starting. We will have much more to say about motors in the next chapter, but for now, it will suffice to note that as a motor ages, it will draw an increasing amount of current. This is because the insulation begins to break down. As the process continues, there is more heat and the deterioration accelerates until motor failure occurs. Large motors in a commercial or industrial setting should be monitored on a regular basis. Either the temperature or current while running at full load should be measured and the results posted on a log so that a damaging trend may be spotted before there is an outage and expensive down time.

Clamp-on ammeters with the hold feature are useful in many ways. If you have three of them, you can leave them hooked up at a three-phase service to see how well the loads are balanced over a time span.

Ground electrode conductors can be checked to make sure that excessive current is not being faulted.

The instrument is valuable in troubleshooting a submersible water pump system prior to making a decision to pull the pump.

There are many other specialized meters, especially for power quality measurements and low-voltage network certification.

The oscilloscope, especially when used in conjunction with a function (signal) generator, allows the technician to see into the operation of electronic equipment in a specialized way.

An oscilloscope draws a graph of voltage against time. Voltage will be on the vertical or *y*-axis and time is on the horizontal or *x*-axis.

The voltmeter will let you know the ac or dc voltage at a terminal, with respect to ground, but the oscilloscope provides much more information. It displays a picture of the waveform for one or more cycles. If the correct waveform is present, you will know that the equipment is functioning correctly at that point and at that time. One way to do this is to have a healthy specimen of the equipment on your bench next to the one that is being repaired. You could compare signals at various points and thereby find the defective stage, connection, or component. However, it is not realistic to think that you would have a good piece on hand. Alternately, other approaches will work.

Some schematics include small graphics that depict the correct waveform at various points. Many of these are available on the Internet. A Sylvania Srt 2232x color TV service manual is readily available there and it shows 16 waveforms as they should be observed at key points.

Such information is available for all sorts of electrical equipment. If you cannot find exactly what you want, you may find the information for a similar piece of equipment.

More Advanced Testing

For a stereo receiver or amplifier, if one channel is not working, you can take readings at various points on the good channel and use that information as a guide. Moreover, if you know the purpose of a stage, such as providing video synchronization, you should be able to figure out what the waveform should look like. Every time you observe a good waveform, you learn something, and if you care to keep a record of readings on each job, you will definitely become adept.

This presupposes that you can operate the oscilloscope. The problem is that waveforms cannot be displayed as picked up by the probe. If you tried to do that, the waveforms would flash across the screen in an indistinct blur. For the waveform to display, it is necessary for the instrument to trigger properly. There are many adjustments on the front panel, and they have to be set properly to get a good display.

More PCB Repair Techniques

There are other concerns as well. For high-frequency signals, probe capacitance becomes critical, and there is the whole matter of impedance matching so that you do not have a chaotic mix of harmful reflections. There are many variations in the ways that an oscilloscope has to be set up, depending on the make and model and the intended use.

The key to working successfully with an oscilloscope is spending a lot of time with the owner's manual so you know how to key in various voltages and frequencies. Many operators' manuals are available free on the Internet, so you can look them over in advance of purchasing an instrument.

The oscilloscope should be dual trace, meaning that it is capable of displaying two inputs simultaneously so that they can be compared. The frequency response should be high enough for the use intended.

The sad reality is that a good oscilloscope is enormously expensive. The model you would want will likely cost more than $5000. There are plenty of old scopes on the used market for under $100, but they are useless. If they work at all, they will be far out of calibration. Servicing and parts are no longer available.

If you work in a facility that has a good oscilloscope and function generator, that is a stroke of good luck and, in time, you can become an expert.

Now that we have looked at some common electrical components, their failure modes, and instruments used to test them, we can examine in more detail the troubleshooting process for commercial and industrial equipment.

First, we have to get past the throwaway mentality that says it is not economically feasible to repair equipment—that once a piece of equipment has malfunctioned it is time to scrap it and buy new. This is in part an inevitable trend, and yet not at all healthy in my view. It is not good for the environment, the economy, or our technological development as a world community.

In the nineteenth century, even as the age of electricity was dawning and industrialization permeating the land, North America was still primarily agricultural. Generations of young people grew up learning to keep steam tractors and sawmills running. This was the setting for a great age of experimentation and invention. Will we lose all of that?

It is sometimes stated that nothing is worth repairing—scrap it and buy new. For small residential appliances and home electronics, that may be appropriate. However, in a commercial or industrial facility, where there is a maintenance department with skilled and knowledgeable personnel, a very different agenda makes sense. A well-managed parts inventory and abundant documentation, together with burgeoning replacement costs, mean that equipment should be kept in service much longer if not indefinitely, given adequate preventive maintenance and informed repair procedures. That approach is advocated in this book.

Residential equipment is tightly engineered, with plastic enclosures that snap together in strange ways. When some of them are opened, parts fly in many directions, or at least expand in a disorganized way so that it is difficult to get them back in place after the repair has been made. When you succeed in snapping the enclosure back together, there is the potential for pinching a wire or worse.

Commercial equipment, while more complex and having extensive subsystems, is actually more amenable to preventive maintenance and repair. A huge washer or dryer in a hotel or commercial laundry is easier to maintain and repair than its household counterpart. Access panels are fastened in place by a large circle of machine bolts that can be quickly removed using your cordless tool. Then you can practically walk inside and look around. Moreover, there is usually adequate documentation so that you can examine the schematic and troubleshooting guide before diving in.

So where does this leave us? The basic troubleshooting steps are:

- Interview the operator.
- Check the incoming power supply.
- Examine the equipment visually, and check for abnormal sounds and smells.
- Based on the symptoms, isolate the defective stage and start taking measurements including testing individual components.
- Still at a loss? Many manufacturers have a toll-free tech support line, and it is likely to help you identify the faulty component. If all else fails, try resoldering key joints and sliding out and in ribbon connectors to repolish the contacts.

This book contains chapters on specific types of equipment. This material should help in troubleshooting them and provide insight into other, sometimes unrelated, equipment. At all times, you should try to develop insight and intuition, until you get to the point where you can instantly see where defective components will most likely be found.

First, we will look at repair methods and materials.

In the event that it is necessary to access the inside of a machine, you will want to take a critical look at the enclosure.

As mentioned earlier, some equipment is notoriously difficult to disassemble. Earlier Apple laptop computers, for example, were quite easy to service. Battery and hard drive were readily accessible behind simple access covers, and they could be easily removed if changes were necessary. The latest version, in the interest of an ever-slimmer profile, has everything including the battery miniaturized and cemented in place in a most user-unfriendly fashion. Is it possible to service this type of equipment? Absolutely. It is just a question of finding the right approach. A wonderful resource is YouTube.com. Type the make and model into the search bar and you will most likely get a how-to video that will demonstrate the best method for opening the reluctant model, plus troubleshooting and repair information. Another good Internet site with a lot of free information is ifixit.com.

If none of the resources we have mentioned is helpful, you must rely on experience and intuition. There will be situations where you just stare at a piece of equipment and cannot see a way in. However, there has to be an answer. Sometimes paper stickers will cover screw heads holding a chassis together. The purpose is to provide evidence of tampering in order to void the guarantee. (If a unit is still under guarantee, that is usually the way to go anyway.)

Another frequent scenario is that screws are recessed above rubber feet, which have to be removed.

> Using a table saw, make a V-block from a 24-in. 2 × 2 and mount it on a short-legged sawhorse. This greatly aids in cutting conduit in the field where a bench vise is not available.

I once had to delve into an older Acer laptop, and could find no obvious way to open the case. Finally, I found the exact model the subject of an electronics technicians' Internet chat room and I learned that the secret was to slide a sharp pointed tool alongside a certain numeral on the keyboard, whereupon a latch inside was released and the case split nicely. Access to various parts on the inside was also difficult, bringing new meaning to the word "refractory."

Often accessing the circuitry inside is more difficult than the actual repair. Afterward, it is necessary to put the whole thing back together without scratching the surface or damaging the latching mechanism. It is also essential to keep track of all the small parts and make sure they go back together as intended. An egg carton or a plastic tray with dividers works well. Start filling sections at one end and progress to the other end. For reassembly, reverse the order.

Often outside cabinet screws have a different finish or a different type of head so that they may be easily distinguished. A major problem with screws is that some of them may be different lengths, and not without reason. If you use a longer screw in the wrong place, you can damage a component or ground out a terminal. Therefore, make notes to yourself or mark the chassis with a felt tip pen, whatever works.

With a little experience, you will find disassembly/reassembly operations much easier.

As far as taking out components and replacing them, what is needed is good soldering skills. Fortunately, soldering is easy. The right tools and materials, some basic knowledge, and a very few hours of practice are needed.

Prior to the 1960s, electrical connections for wiring in buildings were soldered. Electricians twisted the wires together (making sure the copper surfaces were clean), and then applied flux, solder, and heat. Afterward, the joints were wrapped first in electrical tape, and then friction tape to guard against abrasion. The connections were low impedance and reliable, but the process was time-consuming and the taped joints were bulky, making for excessive box fill.

An Essential Skill

A far better way was developed. Solderless connectors ("wire nuts") make for a more efficient workflow and, if properly deployed, result in quality connections every time. An additional advantage is that for testing or alterations, the wire nut can be removed and replaced easily.

Soldering is still essential in electrical work although it is used less frequently. If you need to make a splice where space is limited, such as inside an appliance or hand tool, soldering is the way to go. Printed circuit board components are soldered in place when first manufactured and for repairs. In motor enclosures where the leads have been cut too short or where there could be excessive vibration, soldering is appropriate. Stranded wires that must go under a screw terminal may be tinned to good effect, and inside old light fixtures that are being rewired, soldering is sometimes the best answer. Crimp-on connections for high-current applications (such as three-horsepower and over submersible pumps) gain conductivity and reliability by having solder run into the crimps on either side of the slide-in connectors.

Soldering differs from welding, where the metals to be joined are heated sufficiently to melt to some depth to achieve the good penetration required for great strength. A soldered joint usually consists of two copper surfaces with solder adhering to the surfaces and absorbed below the surfaces to a very short depth. The copper is not melted. Not all metals are equally suited to this process. Copper works very well. Stainless steel cannot be soldered. As for brass, it depends upon the alloy. Some brass solders just like copper, while some is not suitable at all for soldering. Aluminum can only be soldered using specialized materials and techniques. One of the problems with aluminum is that you can get a very nice looking joint that is not good at all. Not all of this is a problem, however, because the only soldering you will ever need to do, most likely, is copper to copper. It is worth noting that the NEC prohibits solder joints in certain applications, notably ground electrode conductors, where a heavy surge could melt the connection.

When soldering wires, they must first be twisted or looped together to make a strong joint that cannot wiggle. Then the soldering process adds strength and conductivity. For a solder joint to succeed, the right materials must be chosen, along with the correct technique, appropriate to the scale of the work.

As for materials, flux must always be applied, prior to making any solder joint. When heat is applied, or after sitting unused for a period of time, any solderable metal including the soldering iron tip will acquire an oxide coating, which comes from oxygen in the air especially in the presence of moisture. This happens instantly when the temperature gets close to the melting point of solder. This oxide layer is very visible. There is a dull or dark appearance to the metal to be soldered and to the soldering iron tip. Moreover, where there is oxide, any attempt at soldering will be unsuccessful. This is because the oxide acts as a thermal barrier or insulation, preventing the heat from getting where it needs to go. Typically, if the soldering iron tip and one or both of the metals to be soldered is oxidized, the solder that you feed in will just crumble and fall away no matter how much heat you apply, the oxide layer becoming thicker and more intrusive all the while. If there is an excessive amount of oxidation, start by cleaning the surfaces with steel wool or scraping them with a knife.

The soldering tip is prepared by tinning it. This involves bringing up the temperature and wiping it on a damp sponge while hot. Give it a couple of swipes on each side, being sure to get the edges and the end. Then quickly, before more oxide has a chance to form, apply a little flux and melt on a small amount of solder. Allow the iron to cool, adding more solder while it will still melt. You should see a nice silvery coating on the tip, indicating that it has been tinned. This is how it has to look prior to any soldering job. The tip usually has to be retinned from time to time, and this should be done after the final usage of the day so that the tip will not oxidize while sitting idle. Remember that oxidation takes place instantly at high temperature and more slowly at room temperature. As long as the tipped is tinned, it will not oxidize.

The correct flux must be chosen. Acid-based flux used by plumbers and for automotive radiator work is too strong, and will reoxidize (corrode) the work after the fact. Moreover, the flux must be nonconductive so that adjacent traces on a printed circuit board or closely spaced electrical terminals will not be shorted out. Furthermore, the flux must not leave a paste residue that would attract and retain dirt that might be corrosive or conductive, or act as thermal insulation. The wrong flux will appear to work at first but is likely to make problems in the future. Buy and stock a resin-based flux that is labeled for electrical work, never acid based. Some solder has a resin flux core that is dispensed onto the work as the solder is melted. This works fine, or you can put the flux on first and use solid solder, whichever you prefer.

Precleaning with steel wool or a knife is useful for getting rid of a heavy coat of oxide, paint, or any other contaminant, but in all cases the flux is essential. Without flux, you cannot solder. Avoid the use of sandpaper as this may leave grains of abrasive embedded in the metal, which would create dead spots in the solder joint, high impedance, heat, and eventual degradation of the connection.

As for solder, there are two variables—wire thickness and alloy composition. You should have several rolls of solder on hand, with different thicknesses for various sizes of work.

If the wire is too thin, it will take a long time to complete the operation and in the process, a harmful temperature may be reached. If the wire is too thick, the large amount of solder will overwhelm the joint, spilling over into unwanted areas and again requiring excessive amounts of heat. With experience, you will choose the right wire size for the task.

Solder is available as alloys of various metals and in differing percentages. They all work to join the metals, but since the melting points vary, the right choice must be made. Alloys with higher melting points are generally more difficult to work with because more heat has to be applied, meaning that nearby sensitive components are at risk. SN60 and SN63, with 40 and 37 percent lead, respectively, are commonly used for electrical work. Lead-free solder is used in plumbing for potable water systems, but the higher melting point makes for problems in electrical work.

In addition, the correct tip size and heat output must be chosen, appropriate to the task. A solder gun (these come in a range of sizes) may be chosen for soldering 14 AWG and larger wires, whereas pencil-tip irons (of various sizes) are right for printed circuit boards (see Figs. 1-30 and 1-31).

If you have chosen the right iron, flux, and solder, and have figured out how to prevent oxide contamination, the soldering job will come together nicely.

Let us start by joining two wires.

After the wires are cleaned and twisted or looped together, suspend and brace them so that the joint is in free air, away from any metal that would pull out the heat and not on wood or any material that would heat up and contaminate the solder joint. Bring the clean, well-tinned iron tip up to temperature while it is in contact with the wires. If they are of unequal sizes, apply more heat to the more massive of the two, so that they reach the

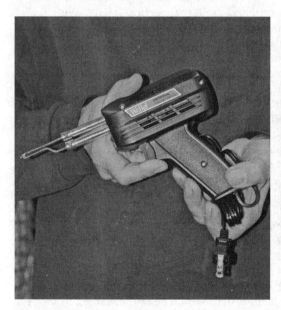

FIGURE 1-30 Gun-type soldering iron. (*Courtesy of Judith Howcroft.*)

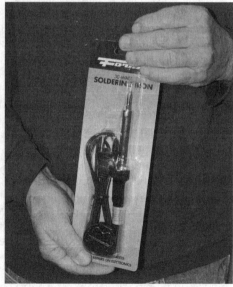

FIGURE 1-31 Pencil-tip soldering iron. (*Courtesy of Judith Howcroft.*)

melting point of solder at the same time. If possible, it is better to apply heat to the underside of the joint. Avoid breathing smoke or fumes because you never know what impurities the metals could contain.

When you begin to approach the critical temperature, touch the end of the solder wire to the joint, preferably on the side of the joint opposite from the source of heat. The solder will serve as an indicator so that you know when the soldering temperature has been attained.

At this point, the soldering iron and solder should be held at angles so that they do not impede your view of the joint. Abruptly, the solder will melt and flow wherever there is heat and flux, even uphill. A poor joint would result from insufficient heat as well as from too much heat, which would start to burn the flux and reoxidize the metal. Therefore, good judgment is necessary as far as knowing when to remove the heat. If you apply solder to the cold side of the joint (away from the heat), wait until the solder melts, and then hold the heat another two or three seconds; you should have a good joint.

After you remove the heat, be sure the joint is not moved until the solder has thoroughly solidified. If it is allowed to wiggle while in an intermediate state, the joint will have fractures. Do not do anything foolish like spraying water or blowing on the joint to hasten cooling. That would crystallize the solder, making for a weak joint.

That is just about it as far as soldering two wires or, for that matter, any two pieces of solderable metal. Practice on scraps of copper. After they are soldered, try to break them apart. Strive for nicely proportioned solder joints, not overly large but feathering out at the edges rather than terminating as big blobs. The result should not have the appearance of a cluster of wild grapes.

Printed circuit board work is based upon the same physical principles. It is a little more sensitive to error and, of course, there is the potential for ruining an expensive circuit board. For this reason, it is a good idea to find a discard and use it for practice. Try removing components and replacing them. Take resistance readings, before and after. Do not forget to use heat sinks and limit the amount of heat used, especially around semiconductors. While active devices are out of circuit, do meter tests as described earlier.

After melting connections and pulling a device, often the hole in the printed circuit board remains plugged with solder. There are tools designed to deal with this problem. While the solder is in a molten state, it can be removed by using a solder sucker, which has a hand-operated squeeze bulb. In addition, fine-braided, resin-impregnated copper solder wicks work well when heated in contact with the plugged hole. You can use these tools in tandem, first the solder sucker and then the braid. Finally, if the hole will not clear, the correct-sized drill bit inserted in a handheld bit holder would work, but beware of metallic grindings that could stick to the board and short out close-spaced traces. Again, excessive heat is ruinous. Besides damaging sensitive components, too much heat can delaminate the board or lift traces.

We mentioned earlier that IC removal and replacement could be problematic. This is because all pins have to be heated and melted out simultaneously so that the component can be pulled free of the board. So much heat would surely damage semiconductors or the board itself. There are ways to work around this problem. You can heat the pins one at a time, then with repeated solder suction and wicking, free up the component. There is such a thing as an IC extraction tool, which can be helpful. This would be if you wanted to save the IC, taking it out of circuit for testing. If you knew the IC was bad from circuit analysis or if it runs abnormally hot to touch, and if you want to save the board, here is an alternate method that works even with a 40-pin DIP.

Using a small, sharp pair of side-cutting diagonal pliers, cut each pin close to the IC body. Then you can melt out and pull each pin individually. A soldering iron with a magnetic tip is helpful. After the holes are cleaned out, the replacement IC or an IC socket can be installed. Alternatively, you can leave the pin stubs in place and solder the new IC directly to them, provided there would be enough vertical clearance when everything is put back together.

There are other circuit board repairs besides removing and replacing components. A frequent repair involves a cracked circuit board. The first thing to ask is why did it crack? If there was traumatic damage to the overall piece of equipment, all stages must be checked out to make sure that some latent weakness does not cause a problem in the future. The cause of a cracked board could be that it became dismounted from the chassis. If it is held in place by screws, make sure they are all in place and tight. The board may go into a slot, and an edge could be damaged. On the other hand, the crack could be the result of age and brittleness, perhaps due to repeated heat cycles that did not reach sensitive components.

After the underlying cause is identified, evaluated, and remedied, the actual repair has to be made. Superglue epoxy works well for repairing broken boards. In addition, you might try any of the Krazy Glue varieties. Plastic glues generally work as solvents, dissolving the parted surfaces so that the materials can mix, fuse, and harden. Cold-press wood glue, which works by seeping into the wood pores and then hardening while clamped, will never work on printed circuit boards, nor would a plain epoxy, which is at its best when there is a space to be filled. A construction adhesive would also not be suitable.

After each stage of the repair, thoroughly clean the board with isopropyl alcohol, finally wiping the surface with a tack-free cloth.

Another method that can be used in addition to cementing the cracked board is to drill small holes on either side of the crack and to insert staples. Flip the board and bend over the ends. Of course, these holes and staples have to be in areas where there are no conductive traces.

If the crack ends in the board without going through to the edge, it is necessary to drill a small stop hole centered on the crack and just beyond where it ends, so that the fissure will not spread in the future. These holes have to be away from traces.

After the board has been repaired structurally, it is necessary to restore any traces that have been broken, nicked, weakened, or otherwise compromised. To start, gently scrape away the insulating solder mask, exposing but not affecting the copper trace. Then, if the damage was at one point only, apply liquid resin flux and resolder with a fine tip, being careful to avoid bridging to an adjacent trace.

If the damage to the trace is for any length, cut out the damaged part and lift it away from the board, using tweezers. Replacement trace material is available from your electronics supply house. Choose a piece the correct width, cut it to length, cement it in place, and solder it in place at both ends. Finally, recoat the area with a nonconductive coating.

An alternate method is to solder a segment of point-to-point wire of appropriate ampacity to the terminals at either end of the broken trace. Whatever you do, where there are high-frequency circuits, you have to be careful not to alter the conductor size or routing as this could alter the characteristic impedance (which must match input and output impedances) thereby causing harmful reflection and data loss or unintended capacitive coupling.

It is hoped that the ideas in this introductory chapter will get you started, at least to the point where you can learn by doing. To diagnose and repair electrical equipment, it helps to know how it works. In the chapters that follow, we will look at a variety of different types of equipment and larger systems. After gaining some experience by reading about and working with them, you should be able to extrapolate into previously unknown areas.

Electric Motors

It is commonly held that a large, 3-horsepower motor is worth rebuilding, whereas a small, fractional horsepower motor is not worth fixing and, if it quits running, should be discarded without further ado. This is generally true, but there are exceptions.

Large commercial and industrial facilities typically have many fans for air circulation in interior areas, to cool transformer rooms, in conjunction with refrigeration equipment, and so on. These fans are usually powered by fractional horsepower, single-phase, 120- or 240-volt induction motors. Quite often workers will inform a maintenance electrician that a fan is not working. It is true that these small motors are not worth rewinding, but there are often simple repairs that provide new life.

Easy Motor Repairs

Frequently, a motor will fail to start when powered up after an idle period. It may or may not make an audible hum, or it can trip the overcurrent device. To begin, check to see if there is power and if the guard or a fan blade has become bent so that the fan cannot turn. (Watch your fingers!) Remember that when the blades are forcing air forward, the fan is pushed back (if there is endplay), so it might be necessary to loosen the setscrew and shift the blade a slight amount. Then, with the power disconnected, try turning the blade. It should be freewheeling, so that if you give it a spin, it will coast awhile before coming to a stop. If it shows resistance, the bearings are probably dry. This often happens with motors that are only about a year old because the bearings are still tight and particularly prone to seizure. The remedy is simple. Spray a little penetrating oil around the bearing where the shaft exits and spin the fan by hand until it is freewheeling. See if there is endplay. If so, work the shaft in and out as this will help distribute the lubricant.

Penetrating oil is good for freeing up seized bearings, but it does not last, so it is best to follow it up with some high-quality machine oil or a little automotive oil—30-weight for hot environments, 20-weight or multi-viscosity if the motor will have to start when the ambient temperature is expected to drop below 60°F. It is usually held that sealed bearings cannot be lubricated. This is not entirely true. There are specialized syringes for injecting oil. In addition, it is sometimes possible to spray penetrating oil around the seal. Some of it will seep in and free up the bearing.

Do not over-oil. Excess oil can spread around inside the motor and deteriorate sensitive parts such as brushes and insulation, and encourage dirt accumulation. Often the front bearing can be lubricated from the outside, but to access the rear bearing will require motor disassembly. However, the front bearing has to work harder, and it is often sufficient just to lubricate that one. It is amazing how often this simple repair works for a motor that was reported "burnt out."

Put the motor back in service and add another drop of oil after the first day's use. Older motors have grease fittings or oil cups. Recommended lubrication interval is given on the nameplate and should be observed.

Another easy fix applies to motors that power woodworking and other equipment in dusty environments, where mechanical parts such as switches and the like may become clogged and fail to operate. Clean them out and you should be good to go. It may be possible to make a shield that will prevent recurrence, but beware of doing anything that will impede airflow.

A word of caution: Certain outside and inside areas are classified according to type and degree of hazard. You should not work on a motor or any other part of the electrical installation unless you have specific training and experience in these matters. One spark now or in the future could cause an explosion or flash fire with tragic consequences.

Before covering motor troubleshooting and repair in detail, we will look at various types of motors as well as their close relatives—generators and alternators.

Motors are generally understood to mean devices that convert electrical energy into rotary motion that can be coupled to a load to perform useful work. Linear actuators such as the common automotive starter solenoid are also motors, and they may fail as well. All motors are subject to troubleshooting and repair, with the exception of repairing some submersible pump motors and the like, where windings are encapsulated, as well as hermetically sealed refrigeration motor-compressors, yet even these have some external wiring that can be checked and repaired.

Two Parts of All Motors

Rotary electric motors are made up of stators, consisting of permanent magnets or windings that are mounted on the inside of the case, and rotors, consisting of windings, permanent magnets, or magnetically permeable soft iron. These elements are attached to a shaft that is mounted in bearings so that it is free to turn. Extending to the outside through an end bell housing is an output shaft for attaching a pulley, gear, saw blade, grinding wheel, or other useful tool. Some motors have output shafts at each end, while some have no output shaft if the work to be done is internal, as in a rotary inverter. Some motors are integral with the driven equipment, such as a refrigerant motor-compressor or a submersible pump. They may be separable.

What makes a motor turn? It is the magnetic interaction of stator and rotor. The current to one or the other has to be switched continually so that the rotor is always chasing after the stator. Therefore, the switching has to be coordinated with the turning, or the turning has to be coordinated with the switching. This switching action may take place inside the motor or outside as in a synchronous motor, driven or regulated by the ac line power with its changing polarity, or it may be a stepper motor, driven and regulated by pulses created for the purpose within the controller.

The basis for all this is the interaction of electricity and magnetism. Where there is electricity there is magnetism and where there is magnetism and a conductor in place, there is the potential for electricity. To have current flow, there has to be relative motion between the conductor and the magnetic field. This relative motion can consist of the conductor and the magnet moving, or of the magnetic field fluctuating.

The relevant formulas together with our still incomplete understanding of magnetism are complex, but everything we need to know in order to troubleshoot and repair electrical motors is straightforward. In the first place, we must see that electrical current and magnetic flux both go in circuits, which in many ways are similar. (Either of them may be an open circuit.)

These are the similarities:

- Magnetomotive force (mmf) in a magnetic circuit resembles electromotive force (emf) in an electrical circuit.

- Flux in a magnetic circuit resembles current in an electrical circuit.

- Reluctance in a magnetic circuit resembles resistance in an electrical circuit.

- Permeance, the reciprocal of reluctance in a magnetic circuit, resembles conductance, the reciprocal of resistance in an electrical circuit.

- Flux is established between the north pole and the south pole in a magnetic circuit, just as current flows from the negative pole to the positive pole in an electrical circuit. (Current is usually thought of as starting and ending at a common source whereas flux exists between two poles that are physically separate.)

- In a magnetic circuit, energy is needed to establish the flux, but not to maintain it. In an electrical circuit, continuous new energy is needed to maintain the flow of current whether it is ac or dc.

The unit of MMF is the ampere-turn, which is created by 1 ampere of electrical current flowing in a single turn of wire. Accordingly, the strength of a magnetic field can be increased by winding multiple turns of the conductor, forming a coil around a magnetically permeable core composed of a material such as soft iron.

We are familiar with Ohm's Law for electrical circuits:

$$E = IR$$

where E = volts (electromagnetic force)
$\quad I$ = amperes (intensity)
$\quad R$ = ohms (resistance)

The similar formula for magnetic circuits is Hopkinson's Law:

$$F = \Phi R$$

where F = magnetomotive force
$\quad \Phi$ = magnetic flux
$\quad R$ = magnetic reluctance

In an electrical circuit, electrons flow through the conductor, which, because it has some slight resistance, dissipates energy in the form of heat. In a magnetic circuit, nothing flows and nothing is dissipated. Magnetic circuits are not linear. Unlike resistance, which is constant, reluctance varies depending upon the strength of the magnetic field and the amount of flux.

We have mentioned that both the rotor and the stator must have magnetic properties. In the case of the stator, this is simple—just wire the electrical supply to permanently mounted coils, which may be connected in parallel or in series. As for the rotor, it is more of a problem. If the rotor were to be wired directly, the wires would quickly twist and break off. Transfer of power to the rotor may be achieved by having brushes, which contact the turning commutator. It, in turn, is wired to the rotor-mounted coils, which have soft iron cores. The brushes are both electrically and mechanically active, so they are subject to failure. Fortunately, they are inexpensive and easy to change. For portable electric saws, handheld electric drills, and similar tools and appliances, they are often accessible from the outside and the only tool needed is a screwdriver. They are held firmly (not tightly) against the commutator by spring pressure. If this spring loses its tension, it may be stretched a little.

An ohmmeter should show continuity from one side of the attachment plug, through the switch (turned on), in one brush, through the commutator and rotor circuit, out the other brush, and back to the other side of the attachment plug. Sometimes, in an electric drill, the switch is double-pole and both sides have to be working. This simple test checks the complete circuit of the hand tool and many other similar items. Leaving the ohmmeter hooked up, you can flex the cord, wiggle the switch sideways, put pressure on the brushes, etc. to bring out and identify intermittent faults (see Fig. 2-1).

There are brushless strategies for getting power into the rotor or not getting power into it but having it active magnetically. Keep that in mind as a primary issue in motor design and repair. If you want to troubleshoot a piece of electrical equipment, especially a motor, you have to know the current flow.

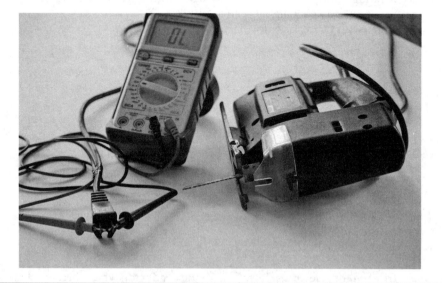

Figure 2-1 To find a fault inside a tool, connect your ohmmeter to the plug. Then try the switch, rotate the armature, move wires around inside, and watch the meter.

The principal types of motors are as follows.

The dc motor was the first motor ever built, long before Tesla and Westinghouse conceived of alternating current and built a generation and distribution system to make possible its use. Even today, there are many dc motors. They were used in large power applications such as elevators and ski lifts, where smooth operation, variable speed, and reversibility are important, before the advent of the variable speed drive (VSD). They are still common in low-power applications including automotive starters, ink jet printers, children's battery-operated toys, and tools and appliances for off-grid users that have dc power.

DC is supplied to the stator coils, creating a stationary magnetic field. Current is also fed to the rotor. Unlike the stator, this field cannot be stationary or else the motor would rotate at best a part turn, finding the place of least reluctance. To make the dc motor work, the rotor current and resultant magnetic field have to be switched regularly, and this is done by the commutator, which performs its function in time with the turning of the rotor, so that speed may be varied merely by adjusting the supply voltage, and direction may be changed by reversing polarity. DC may be supplied from a battery, dc generator, solar PV system, or utility power that is rectified.

There are two broad categories of ac motors—synchronous and induction. By their very nature, neither of these will work on dc.

A synchronous motor works on the principle of reversing magnetic fields, but the switching comes from outside. It is inherent in the nature of the power supply, whether utility, on-site ac generator, or dc changed to ac by means of an inverter. Rotation is dependent upon line frequency, generally 60 Hz in North America. The period of rotation is equal to some multiple of line frequency, based on the number of magnetic poles. Speed is precisely keyed to the frequency and will not be altered by adjusting the voltage.

Because of this frequency dependence, synchronous motors will not start, except for very small motors where the inertia of rest is easily overcome. Large synchronous motors have separate dc excitation, or they are brought up to speed by a small auxiliary motor.

Large synchronous motors are noted for their superior efficiency, and with a favorable power factor, they help to balance out equipment with a lagging power factor significantly reducing energy costs for a large industrial occupancy.

> The National Electrical Code requires emergency lights to be regularly inspected and a record kept. The time interval is not specified. In a large facility, maintenance electricians are entrusted with this important job. Place the results in an Excel file and compare results to see which units are eating batteries. They will need new circuit boards.

Induction motors are much more common, and the method for getting energy into the rotor is a brushless and overall quite elegant solution. They are called "asynchronous" because the speed is not directly synchronized to line frequency, although it is dependent upon it. They require ac to operate. Induction motors are also called "squirrel cage" motors. The principle is that the power gets into the rotor by means of electromagnetic induction. In effect, the stator is the primary and the rotor is the secondary of a power transformer, so no brushes are required.

Slip in an Induction Motor

In an induction motor, the rotor speed is *not* synchronized with respect to the stator's rotating magnetic field. If it were, there would be no relative motion, no induction, and no power transfer. The rotor speed is a certain percentage less than the stator's rotating magnetic field. This difference is called "slip" and it is what makes an induction motor possible. It is important to realize that in an induction motor, the rotor is not merely out of phase with the stator's rotating magnetic field, but it is actually turning more slowly. Increasing the frequency will increase the speed, but there is no direct synchronization.

The rotor windings are closed loops. Since power is transferred by magnetic induction instead of brushes, induction motors are fairly maintenance free.

The typical amount of slip for a small induction motor is around 6 percent. For a large induction motor, it will be about 2 percent. If, due to overloading, the slip increases beyond 20 percent, the motor will stall and if power is not disconnected soon, the motor will overheat, the windings will become heating elements, and the motor will be destroyed.

A stepper motor is a brushless dc motor that, unlike other motors, is able to turn in small increments, fractions of a rotation, rather than continuously. The amount of each increment is called the step angle, which varies depending upon the stepper motor model. It can be less than 1 degree or as much as 90 degrees, or a quarter turn. The stepper motor turns one step, then stops whether or not voltage is removed from the circuit that feeds the motor. If current is maintained through the winding, the motor remains at rest but maintains a definite holding torque—a very useful function in many applications (see Fig. 2-2).

If you apply a steady dc voltage to one pair of wires, the motor may or may not turn one step, depending on the polarity of the voltage, which pair of wires the voltage is applied to, and the position of the rotor. If it does turn, it will turn only one step and then stop, holding that position with available torque. It will remain motionless until another pulse is applied.

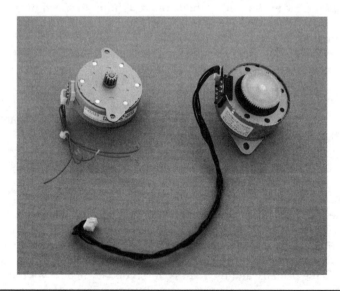

Figure 2-2   Stepper motors from ink-jet printers.

Accordingly, it may be said that the commutation takes place outside the motor, and is the province of an outside controller. Depending upon the characteristics of the motor, an audio frequency may be applied and the motor will really spin, albeit rpm levels are generally less than for other motor types.

Since all commutation takes place outside, stepper motors have no brushes or moving parts to wear out, other than the spinning rotor. For this reason, all stepper motors are extraordinarily long lasting and reliable. Failure modes are bearing fatigue, which is infrequent because of low rpm and duty cycle when operated in stepping mode together with typically light loading, and overvoltage on the field coils, which can be mitigated by installing overcurrent protection and protective diodes to prevent inductive voltage spikes.

To understand how stepper motors work, it is instructive to consider the various forms they have taken.

The variable reluctance stepper motor is remarkably simple. It has two parts—stator and rotor. The rotor is mounted by means of front and rear bearings so that it is able to rotate and perform useful work by virtue of its output shaft and an attached gear.

A variable reluctance stepper motor is characterized by a distinctive feature—it has a soft iron notched-tooth disc. Soft iron is highly permeable to magnetic flux; that is, it has low reluctance relative to air. That is the reason it is called a variable reluctance motor. The stator is equipped with field windings, usually three, four, or five. This would imply six, eight, or ten wires, but often there is a single common conductor with the connection made inside, so there will be one wire for each coil plus one common wire. These are easy to ring out with an ohmmeter and to label. The common wire is generally connected to the dc power supply positive terminal. The controller applies power to each of the windings in sequence. A series of pulses is what is needed. Like all stepper motors, the variable reluctance version will respond to a wide range of voltages. In many cases, you can manually pulse and step variable reluctance units with a 9-volt battery.

Even without seeing what is inside, it is possible to identify a variable reluctance stepper by the fact that, with no power applied, it will turn relatively freely by hand whereas the permanent-magnet version will exhibit a more pronounced cogging.

In contrast to the variable reluctance stepper motor, the permanent magnet stepper relies upon attraction and repulsion between a rotor-mounted permanent magnet and stator-mounted windings. Both structures have sets of teeth, offset with respect to one another, so that when timed voltage pulses are applied to the stator windings, the rotor will turn in an incremental fashion. The greater the number of teeth, the smaller the step angle, making for a finer resolution.

Permanent magnet stepper motors may be unipolar or bipolar. The bipolar stepper simply has a single winding per phase on the stator. The unipolar stepper is similar, but each winding has a center tap. The unipolar may be used as a bipolar stepper by disregarding the center taps and leaving them disconnected. The bipolar unit is a simpler design, but control becomes more complex. For the unipolar stepper, the magnetic field direction is reversed by switching on each section of windings, thereby reversing direction. A microcontroller will turn on the drive transistors in the correct order. Unipolar steppers are easier and less expensive to implement for this reason.

As noted, bipolar stepper motors have a single winding for each phase. It is necessary to reverse current direction to reverse magnetic poles, making the drive circuit more complex. An H-bridge is required.

Hybrid steppers combine characteristics of variable reluctance and permanent magnet types. They provide superior speed, torque, and step resolution, but at a greater cost. Step angles may be as low as 0.9 degrees, which translates to an incredible 400 steps per revolution. The rotor has numerous teeth and a concentric shaft-mounted, permanent, axially magnetized magnet. The teeth on the rotor mean increased holding and dynamic torque in contrast with variable reluctance and permanent magnet versions.

If it is possible to use the permanent magnet stepper, the cost savings are significant.

For prototyping and experimental purposes, stepper motors may be easily obtained by disassembling inkjet printers. These typically contain two steppers and one 2-wire dc motor, valuable components for robotics projects. Ink-jet printers are usually quite inexpensive because the manufacturers provide them at low cost and then make their money by putting a high price on replacement ink cartridges. Moreover, a new computer frequently comes with a new printer, so ink-jet printers are everywhere and may be had for the asking. Similarly, fax machines, dot-matrix printers, and disc drives have stepper motors. These should be perfectly good long after the overall machines are worn out or obsolete.

Another excellent resource is an old, obsolete, large satellite dish, also freely available. These had larger stepper motors because, unlike modern small satellite TV and Internet access dishes, a directional motor was required to aim at different satellite locations (see Fig. 2-3).

When harvesting stepper motors from obsolete or surplus equipment, it is desirable to save full-length leads with connectors, as well as mounting screws and hardware and any peripheral electronics such as power supplies, LEDs, protective fuses, diodes, drive transistors, and the like.

Small motors that are integral parts of factory-made equipment may not have nameplates. On stepper motors, markings may be incomplete or nonexistent. If there is a name or number stamped on the stepper motor, it may be possible to do an Internet search and print out a datasheet. Another approach is to plug in the printer and take voltage readings.

Figure 2-3   Old-style satellite dish with stepper motor.

Stepper motor applications include:

- 3-D printing
- Telescope clock drives
- Solar array trackers
- Computer peripheral and business machines
- Robotics

In terms of robotics, a number of considerations arise. Steppers must have an appropriate resolution, based on step angle, for the function that is contemplated. There must be sufficient power for the mass involved. A smaller stepper is better in terms of onboard space required and power consumption. For robot mobility, too much torque can mean a tendency to spin or slip, whereas too little torque can result in stalling or inability to climb out of a hole or negotiate an uphill slope. Either of these limiting cases can spell loss of mobility.

A stepper motor is commutated and controlled from outside. To work, it has to be supplied with a brain.

For robotics projects, a great way to implement these functions is by means of an Arduino, which is an open-source electronics prototyping platform based on flexible, easy-to-use hardware and software. An assembled board costs approximately $30. It is a programmable controller with numerous and diverse features. It is cross-platform, which means that you can connect to your Mac or PC using a USB cable. Drivers and software are available free of charge at www.arduino.cc. The site also includes reference material with each Arduino command and relevant code snippets. The Arduino design is open source, so the hardware is available (assembled or in kit form) from various manufacturers.

A rather complete and well-written book on Arduino use including programming and interfacing with robotic projects is *Arduino Robotics*, by John-David Warren et al. This work assumes little prior knowledge, yet takes you far into the world of electronics for robotics.

> When upgrading to a larger service, the old branch circuit wiring must be removed from the entrance panel so that a new breaker box can be installed. Do not pull the stripped cables out through the connectors—that is sure to nick the conductors causing shorts later. Instead, remove locknuts from the inside of the box and leave the connectors in place on the cable jackets. Reattach in the new box. That way is easier and less hazardous.

Returning to motors in general, in troubleshooting motorized electrical equipment, there are two issues to consider: whether the motor or its control circuitry is at fault and, if so, what is the nature of that fault?

A Common Fault

There are instances when the motor fails to turn when the machine is powered up, and yet the motor is good. An example is a large commercial dryer, as found in the laundry of a big hotel. One of these units consists of a large drum capable of holding sheets, tablecloths, towels, clothes, and other items to be dried straight out of the washer. The drum capacity is many times greater than that of a residential-type dryer. These machines require a large motor to turn the drum in order to tumble dry the contents, and a means to inject hot, dry air into the drum and expel the cooler moist air, usually vented to the outdoors.

The heat may be electric, like most residential dryers, gas, or steam, the latter heating air by means of a heat exchanger.

There is a door that is opened to add and remove contents and, like a residential model, there is an interlock switch with a relay to interrupt power to the motor when the door is not firmly closed. This simple switch can be checked with an ohmmeter or by operating it by hand. The switch may be defective (open) or the mechanical linkage that operates the switch in response to the door position may become bent or broken so that the motor will not run.

It is to be emphasized that a broken interlock switch should not be electrically shunted or otherwise defeated just to get the machine running, even on a temporary basis, as severe injury to a worker could occur.

This interlock switch is one of several devices or systems that are in series to control the motor.

Front panel controls include a main power switch and pilot light, provisions for setting the temperature so that it is appropriate for the material to be dried, and a timer calibrated in minutes to regulate the time of the drying cycle and a shorter cooling-down cycle. As mentioned, the door interlock switch cuts power to the motor. It also shuts down the heat, if electric, but if you open the door and quickly look inside, you may be able to see light from the heating element before it fades entirely.

In older models, these controls operate relays to switch the power to the motor, heating elements, and blower. Newer units invariably have a printed circuit board that performs the same functions efficiently and reliably.

If you verify that you have power to the motor and find that it is entirely unresponsive, or if there is no power to the motor because it has tripped the overcurrent device, it is probable that the fault lies within the motor. However, you need to verify that the driven load is not locked up or turning abnormally hard. On moderately sized equipment, it may be possible (with power disconnected and locked out) to turn the driven load by hand. You should be able to judge, depending upon the nature of the equipment, if it has the proper drag. The other thing to do is disconnect the load. If power transmission is via a driveshaft with a universal joint or semiflexible coupling with rubber bushings (a Lovejoy coupling), some wrench work will have it apart. However, when you power up the motor, take care that there is nothing that can whip around or otherwise create a hazardous situation. If the motor still will not run and you have verified continuous correct voltage to the terminals with no tripping out, you have to conclude that the motor is defective. You can go a little further. At times, there is a broken or burnt wire inside the case, and if the break is not too close to the winding, it could be amenable to soldering or to a crimp connector if there is room inside. When making repairs of this sort, it is essential to ensure that there is nothing live that can ground out against the enclosure given that there will be vibration and temperature changes when the motor is running.

Now we come to the important topic of brushes and commutators. They are subject to wear and damage from overheating, but the motor is constructed so that they can easily be replaced. Of course, not all motors have brushes. Brushless (permanent magnet) dc motors, stepper motors, and induction motors are all unencumbered by brushes, but where there are brushes (conventional dc motors, synchronous motors, shaded-pole motors, and universal motors), they are the first things to check.

Early dc motors conveyed electrical current through the commutator and into the armature by means of actual braided copper brushes that were connected to the outside dc power source—hence, the name. Because of their high conductivity and inaccurate contact with the commutator, they have been replaced by graphite, which transitions more smoothly from one commutator bar to the next.

As with many things electrical, heat is the enemy. Regarding brushes, this is definitely true, and the primary source of heat is sparking. Graphite against metal is an excellent lubricant. (If you want powdered graphite to use as a lubricant, grind up some discarded brushes.) As for sparking, in a new motor it is minimal, but eventually the commutator or brushes may become pitted or distorted, and there will be a discernible sparking. Of course, the heat that is generated accelerates this deterioration.

Many motors, such as in portable drills and circular saws, have openings (for ventilation) in the housing, so that it is possible to watch the brushes ride over the turning commutator while the motor is running. If you observe a number of these motors that you encounter, you will quickly get a sense of how much sparking is normal and when it becomes excessive.

Once sparking at the brushes becomes excessive, it rapidly gets worse until it has the appearance of a gas flame. The result will be a damaged commutator if not a burnt up tool or even an electrical fire. Therefore, at the first sign of hot sparking, the brushes should be replaced if new brushes are available. If they are not available, the alternative is to fix the existing brushes if they have not become too short. You can wrap a piece of sandpaper, abrasive side out, around the commutator, press the brushes one at a time against the sandpaper, and rotate the commutator to grind down the brushes a slight amount. This will smooth the mating surface and make it conform better to the curvature of the commutator. Afterward, clean the brushes and commutator with isopropyl alcohol to ensure that no abrasive particles remain. The lengths of the brushes and pressure supplied by the springs should be uniform. Then, if you try running the motor, you should find the sparking greatly diminished.

If replacement brushes are not available, some resourceful individuals have attempted to make new ones by cutting down larger brushes that are on hand, and grinding the correct curvature. This procedure may not provide a long-lasting solution because brush material and the heat treatment ("sintering") that is used to harden it are carefully designed by engineers to match the electrical and mechanical characteristics of the motor.

Going Deeper

As for the commutator, it should be inspected and serviced when necessary, perhaps every second time the brushes are changed. Anything that is not right with the commutator will translate into rapid brush wear, further damage to the commutator, and poor motor performance. The commutator should not be deeply grooved where the brushes ride. The pathway should be polished, but not ground away, and it should not be rough or pitted. Measured with a micrometer, the commutator should not be out-of-round or irregular. Specifications are found in motor manufacturers' service manuals.

The insulating gap between commutator segments must be recessed a slight amount. After the commutator has been cut down on a lathe, it is necessary to deepen the grooves between segments. There is a special tool for doing this, or a hacksaw blade will work.

Overheating, moisture, exposure to corrosive materials, or excessive voltage will damage motor windings. Repairs must be performed by a motor rebuilding shop, and should not be attempted unless you are prepared to invest in the equipment and acquire the expertise that is involved. Such shops have large ovens in which motors are placed before and after repairs are completed. Over a period of hours, moisture is baked out of the windings in order to prevent internal current leakage and other harmful effects.

Other than brushes, the major replacement items in electrical motors are the bearings. These may be sealed or equipped with grease fittings or oil cups so that they can be lubricated. The trend today in motors as well as all kinds of machinery is toward sealed bearings. There is a palpable savings in labor and materials over the life of the machine. In addition, the risk of introducing dirt or abrasive material is eliminated. Nevertheless, many technicians prefer bearings that can be greased, feeling that these are more trustworthy.

However, it is a fact that improper lubrication will result in bearing failure, so it is necessary to follow these guidelines:

- Do not neglect the specified lubrication interval. In high ambient temperatures or when heavily loaded, when the load is pulley driven as opposed to shaft driven, the interval should be shortened. Side loading puts more pressure on the bearing surfaces and tends to squeeze out the lubricant.

- Do not introduce contamination. Before fastening the grease nozzle onto the fitting, remove any dirt accumulation in the area, and then wipe off the fitting with a clean cloth. Avoid setting the grease gun down in an area where it will pick up dirt and, if necessary, wipe off the nozzle before greasing. Any abrasive material that finds its way into the bearing will wear the metal surfaces and cause premature bearing failure.

- Do not overlubricate. Too much grease will burst the grease seals, eventually allowing moisture and contamination and wasting grease. Overlubrication is also harmful, especially in high-speed applications, because when the bearing is overfilled, the moving parts have to work harder, eventually making for overheating. It is better to do more frequent light lubrications.

- Keep records. At the location of a large, expensive motor, a log should be posted with the date and description of any maintenance operations including lubrication.

- Use the correct grease. For low-speed operation at moderate temperatures, a multipurpose lubricant is appropriate. Check the operator's manual if in doubt.

Temperature measurements, taken at regular intervals and entered into the maintenance log, are helpful in keeping track of bearing wear so that bearings may be changed at a time when downtime will not interrupt the work flow. A number of different instruments will measure temperature even at a distance. They are easy to use and accurate.

Motors can be tested using a variety of meters. The tests are made both in and out of circuit. Assuming a motor has been in use and continues to perform in a satisfactory manner, begin a preventive maintenance program. The results should be written up in reports and analyzed on a regular basis so that there will not be an unexpected outage.

To begin, take the motor out of circuit. This can be done without disrupting terminations, cutting wires, or introducing unnecessary splices if there is a proper disconnecting means or overload protection. Many motors can be isolated at the controller. An example is a properly installed submersible well pump. Most makers use the motor and control box manufactured by Franklin Electric, bolting their proprietary submersible pumps to it. Some, like Goulds Pumps, put their own name on the box, but most retain the Franklin branding (see Fig. 2-4).

In a three-wire system, a red wire provides 240-volt starting current, black is for the run winding, yellow is common, and there is no neutral. A bare or green wire is the equipment ground. (It is not a conventional three-wire, two-voltage system as in a common single-phase service.)

The electronic components, capacitor, microchip, relay, etc. are mounted on the cover, so that when it is removed, there is access to these items. Moreover, the pump motor, underground wiring, and conductors going down inside the well casing are isolated from the power source. Often a failed system can be restored merely by replacing the relatively inexpensive cover. The meter readings are documented inside the box. These specifications vary depending upon the horsepower of the motor.

Figure 2-4
All the electronics for a submersible well pump are in this control box.

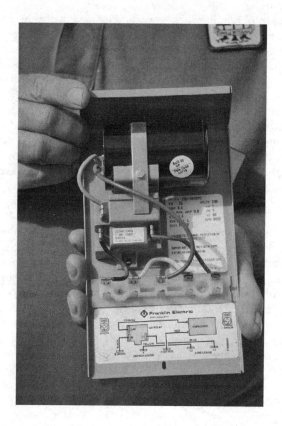

With the cover removed, there is access to input and output terminals. The input terminals are still live with 240-volts from the branch circuit.

Using the multimeter volts function, check the input. There should be 240 volts between the line terminals and 120 volts between each of them and the bare or green equipment ground. Next, switch to the ohms function and take readings at the output. Compare these values to the manufacturer's specifications. Like any motor, the resistance of windings to ground has to be high—in the megohm range. If it is low, the motor windings are grounded, and the motor must be replaced. (A submersible water pump motor is sealed with epoxy-encapsulated windings and cannot be repaired.) Alternatively, there is a fault in the wiring between the control box and the pump. Very often, a wire chafes against the inside of the well casing, or a fault develops in the underground wiring.

The next test also involves ohms readings at the control box output. Check the resistance individually between red and yellow, and between black and yellow. These readings have to be in the low ohms range, as specified in the box, varying with the horsepower. If the resistance is too low, near zero, there is a short. If it is high, you have an open circuit, either in the wiring or in the motor.

Sometimes these faults are found to be in the wiring right at the wellhead, where there are typically wire nut connections, and that is an easy fix.

These resistance tests are the way to get started, but they are not definitive. Information that is more complete is gained by taking current readings with the cover replaced and the motor running.

These current readings are best taken using a clamp-on ammeter. As mentioned previously, the reading has to be taken on only one conductor. If the hot and return wires are both enclosed in the jaws of the clamp-on ammeter, the two currents will cancel because they are flowing in opposite directions, and there will be a reading of zero. Therefore, you have to find a place where the wires are separate in order to take this measurement. If the wires coming out of the enclosure are in raceway (as they should be), that means it is necessary to look elsewhere. Some possibilities are inside the entrance panel or load center, or under the well cap. The current readings must conform to the specifications shown inside the control box, in the owner's manual, or posted on the Internet for the system to work. Of course, if the system is cutting out, either tripping the breaker in the entrance panel or load center, or cycling at the control box, these readings cannot be taken. In addition, if there is an open circuit so that the motor is not running, there will be no current flow. Otherwise, the ampere readings will reveal the condition of the motor.

Submersible pump motors present unique challenges in terms of troubleshooting because access to the motor requires pulling the pump/motor assembly up from the depths.

The test procedure for other types of motors is similar, but usually they are right out in the open. Start with the multimeter volts function to check supply voltages, then take ohms readings with the motor disconnected from the power source and any control wiring, and finally do a dynamic test by taking clamp-on ammeter current readings with the motor running, if possible. This final test is useful when a motor trips out after running under load for a period of time, sometimes 15 minutes, sometimes an hour or more (see Fig. 2-5).

High current draw can mean that the motor is getting tired, that is, it is in need of rebuilding or replacement, although a thorough cleaning and a check of all splices and connections will sometimes solve the problem. Rarely, a breaker or overload protection trips at an abnormally low current level. The clamp-on ammeter reading will clarify the situation.

Figure 2-5
A clamp-on ammeter will reveal the condition of many types of motors.

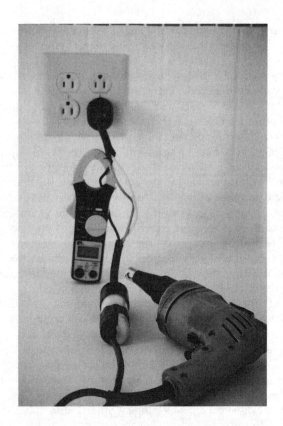

Other Motor Test Procedures

There are other, more advanced, tests that can be made. A spectrum analyzer or oscilloscope will reveal power quality problems—clipping or harmful harmonics—that cause poor motor performance and perhaps overheating of the conductors. These problems can occur when certain other equipment is simultaneously running, an example of the load influencing circuit parameters.

Then we come to the important subject of the megohmmeter, or megger (named after one of the companies that manufactures it; Romex and Amprobe are also trade names that have become generic) as it is called. This valuable instrument can be bought for between $150 and $500. Battery-powered and hand-crank (magneto) versions are available. Since high voltages are used, there are important safety considerations. One hazardous aspect of electric shock is that the muscles involuntarily contract, so that it is possible that the affected individual, grasping an energized object, will be unable to release it. The victim will be exposed to the harmful voltage on a continuous basis, unlike the usual scenario when one recoils and receives only a brief jolt. The skin, normally a partial insulator when it is dry, begins to break down and blister, becoming more conductive and increasing the hazard. This danger to the megger operator is lessened when the hand-crank model is used, as presumably the hand cranking will stop immediately.

Accompanying the megger is an extensive technology that is at once very revealing and a war of nerves because if it is overdone, the object being tested will have its insulation degraded.

The megger is used to test the insulation integrity of many types of electrical equipment, as well as power and data cable before or after installation. One of the underlying ideas is that very high amounts of resistance, such as many megohms, cannot be tested with a conventional ohmmeter because the low test-voltage is insufficient to create a current that is measurable. The other underlying idea is that many materials will present a high resistance to the flow of electrons unless a sufficiently high voltage is applied for a length of time, whereupon an ionized path is established and the electricity blasts through.

A prominent example is lightning. A static charge, due to migration of charged particles from one area to another, builds in intensity until there is a sudden ionization, or breakdown in the insulation that is the air between the two regions. An electrical circuit is established and there is an enormous flow of current for a brief period of time.

When there is a question regarding the insulation integrity of an electrical motor, megger tests are commonly conducted. In research laboratories and for product development, the test can be prolonged to a point where the motor or other equipment is destroyed in the interest of gathering information concerning the underlying design. However, in our maintenance and repair environment, especially when it is a large, expensive motor, we want to make sure that the test remains nondestructive, which means placing a limit on the voltage level and duration of the test.

As time passes, electrical insulation will deteriorate, usually very slowly, but rapidly if there is exposure to moisture in the air or due to flooding, nearby leaking pipes, corrosive vapors, or liquids or physical trauma. Temperature extremes, vibration, or other conditions may be factors. Harmful situations will compromise winding insulation inside a motor, and the megger test can reveal a developing situation before it becomes catastrophic, by which we mean hazardous to productivity, property, or human life.

The megger is a little more complex than a conventional ohmmeter. Besides the internal generator or power source, it consists of both voltage and current coils, so that variations in the amount of applied voltage do not invalidate the resistance readings. The amount of applied voltage is usually between 500 and 1000 volts. (Higher voltages are used in a high pot tester, which requires specialized training and familiarity with what is called medium-voltage work.)

There are several types of megger tests. The most frequently performed is the short time or spot reading. The meter is connected across the insulation to be tested and the voltage is applied for 1 minute. For a motor, an acceptable resistance reading is considered 1 megohm for each 1000 rated volts. For any voltage under 1000, the reading should be at least 1 megohm. This amount is subject to variation, however, because of changes in ambient temperature and humidity. In addition, a new motor will exhibit a much better reading, especially if it has always been stored in a dry location. (Some large, expensive motors have provision for a continuous low-level voltage to be applied to the windings when the motor is not being used, the current resulting in just enough heat to drive out the moisture, thereby preserving the insulation integrity.)

By taking and recording regular readings at periods of uniform temperature and humidity, it is possible to spot trouble on the horizon and deal with it during scheduled downtime as opposed to experiencing disruption to production or worse in the event of catastrophic failure.

The time-resistance test consists of taking successive readings at specified intervals of 5 to 10 minutes. Good winding insulation will exhibit an increased resistance at each reading, while bad winding insulation will show a drop in resistance at each reading. For this test, the focus is not on the exact values, but rather the relative change.

Some meggers have multi-voltage capability, and if you have one of these instruments, you can do a stepped-voltage test. It involves testing winding insulation resistance—first at a lower voltage, and then after discharging any voltage that is stored capacitively, at a higher voltage. If there is a significant drop, you will have reason to believe that the integrity of the winding insulation is compromised. It is possible to apply to the motor moderate heat in a dry location for an extended period of time. There are instances where this procedure will give new life to the motor, improving megger readings and motor performance. More likely, the unit will have to go to a motor rebuilding shop for an extensive overhaul.

It is to be emphasized that megger tests should be performed on a motor only when it is disconnected from the power source to ensure that hazardous voltage is not backfed to wiring, circuits, or equipment elsewhere, invalidating the test results as well as creating outside damage or danger to persons.

Three-phase motors are less expensive and simpler to install, maintain, and service than their single-phase counterparts. They will start on their own; hence, there are no separate start windings, circuitry, wiring, or switch mechanism with which to be concerned. Moreover, it is not practical to build a single-phase motor over 5 horsepower, so they are just about all three phase. Small, fractional horsepower motors also may be three phase, and they are common in industrial locations. Why, then, are not all motors three phase? The answer is that three-phase power is not available at all buildings, either because of the service or because of the utility distribution system.

Driving throughout suburban neighborhoods, it is instructive to examine (visually, from a safe distance only) the distribution lines. Three-phase power is characterized by three hot legs at the highest level, with a grounded neutral a few feet lower. Well below these power lines are telephone, CATV, broadband cables, and so on. Single phase, in contrast, has one hot conductor at the top, with a grounded neutral at a lower level and "low-voltage" communication and data cables below. The principle is that there is to be sufficient separation so that in an aerial bucket workers can service the lower-voltage conductors without coming near the high-tension lines. In addition, sufficient elevation ensures that trucks and equipment will not snag on the lines when passing underneath, and children will not reach them by climbing the poles. Ground clearances and conductor separation specifications are all spelled out in the National Electrical Safety Code, which covers, among other things, utility wiring (excluded from the National Electrical Code).

You will notice where there is three-phase power a single-phase service conductor is connected to one of the three-phase utility lines for each individual single-phase step-down distribution transformer. This is done on a rotating basis in order to keep the loads balanced insofar as possible.

You will see that the three-phase lines follow roads in more populated and industrialized areas, while single-phase lines extend further into more remote areas where electrical usage is lighter.

The key concept is that single-phase service can be derived from a three-phase line by means of simple electrical connections, but three-phase cannot be derived from single-phase power except by means of specialized equipment—a rotary or electronic phase converter. Single phase is derived from three phase at the power pole or inside the building. It is a frequent layout to have three-phase power to a high-rise building, broken down into single-phase circuits emanating from load centers on each floor, again with a view to balancing the loads. No transformers or substations are necessary, just double-pole breakers within three-phase boxes. Three-phase panels have three busbars, so that either type of power is available.

Three-phase motors are installed, maintained, diagnosed, and repaired in much the same way as single-phase motors, with the exception that phase order becomes very important.

Three-phase motor direction is changed by reversing the connections of any two of the three legs. A single-phase motor, with the exception of a universal motor, cannot have shaft rotation reversed merely by switching wires, for there is no polarity distinction in alternating current. A single-phase motor with reverse rotation capability has separate clockwise and counterclockwise windings, with four wires in theory, but actually three because one is common. The three wires are brought to an outside box where the operator controls rotation and start-stop action. A three-phase motor can be controlled in a similar way, although frequently it is intended to run in one direction only. This is done as part of the permanent installation. The conductors may be changed at the motor, a junction box, a load center, or the service-entrance panel, as long as other motors are not on the circuit. So now the question is, how do you ascertain the wiring connections for the desired rotation of a three-phase motor? It is impossible for information on the nameplate or in termination markings (unless they are added after the installation) to provide the answer because there is no defined phase order in the three-phase circuit, not to mention the utility supply.

The easy answer is by trial and error. However, beware that certain motor/pump units can be damaged if run the wrong way, with the seals being instantly destroyed. For many installations where the load will not be negatively affected, it is a simple matter of observing the action. Some pumps and fans will move liquid or air in the correct direction for either motor rotation, but one way produces more output than the other does due to the cup shape of the blades or impellers. This is true of submersible water pumps.

Another method for ascertaining motor rotation is by means of a three-phase motor rotation tester, available for a little over $100.

> If a telephone has no dial tone and appears "dead," many times it is just the line cord, receiver cord, or receiver in that order of probability.

There is another phase connection issue for three-phase motors. The fact is that not all three legs may have precisely the same voltage, due to local single-phase loading and line impedance between transformer and load. At the same time, the motor windings may exhibit slight differences in impedance. Phase imbalance will be either increased or decreased depending on how the lines are matched to the motor windings. The objective is to get uniform voltage with the motor running. To this end, it is possible to change the connections to get all three possible combinations without reversing the motor direction. This is done by moving A to B, B to C, and C to A. The procedure is called "rolling" the connections, and performing this tune-up will often improve motor performance and reduce operating temperature.

Where three-phase motors are in use, reliability and efficiency are prime concerns; and for large, expensive motors that play a key role in the workflow, it is essential to have a good preventive maintenance program in place. The basic elements are temperature monitoring and lubrication—with a written log posted nearby—as well as cleanliness, so that dirt and debris will not accumulate and impede air circulation necessary for cooling.

Additional Considerations

Do not neglect the load. It may not be part of the electrical installation per se, but it impacts motor performance greatly. Any binding or mismatching can cause the motor to be overworked, making for expensive operation and short motor life. Vibration, moisture, and high ambient temperature are harmful for three-phase motors. The two primary concerns, as with single-phase motors, are insulation integrity and bearing life.

Years ago, there were more dc motors, and speed control could be accomplished simply by varying the supply voltage. This is not possible for an ac motor. Some ac motors operate at two or more speeds, but this is accomplished by having separate windings for each speed, with appropriate switching.

Rotary power in industrial settings has been vastly enhanced by the development of the variable frequency drive (VFD). This is an umbrella term for an ac motor that receives its power from a main drive controller in conjunction with an operator interface. The usual ac motors controlled by VFDs are three-phase synchronous and induction machines. While synchronous three-phase motors are sometimes preferable, the three-phase induction motor pairs up nicely with a VFD and has the advantages of lower cost, simplicity, and ease of maintenance. The combination is very workable and it means you can get a continuous speed range out of a single-speed motor, as required in many applications including high-horsepower installations.

Most VFDs conform to a single design, although some are built with numerous bells and whistles for enhanced operation and self-diagnostic capabilities. Typically, a three-phase 460-volt ac input comes into the enclosure via three conductors of the proper ampacity based on the drive's nameplate rating.

A heavily heat-sinked three-phase full-wave diode bridge provides dc to large capacitors. The dc goes into the inverter section, which produces three-phase ac power for the motor. The inverter contains transistors, diodes, and other electronics that create this waveform at the desired frequency to determine the speed of the three-phase synchronous or induction motor.

Troubleshooting the system is greatly facilitated by having on hand the manufacturer's manual with schematic and parts list.

A VFD presents diagnostic challenges but an orderly approach and understanding of the fundamentals will work wonders. As always, begin with a visual inspection. VFDs often operate in a harsh environment, with a lot of vibration, heat, and dust. Moisture is always a possibility. Thoroughly clean the inside of the enclosure and surrounding area after locking out power and discharging all capacitors. A vacuum cleaner is the tool of choice. It should have a plastic nozzle so that nothing can be shorted. Heat sinks must be thoroughly cleaned, as dirt will impede heat transfer. Make sure that there are no foreign objects including manuals or paperwork that could block airflow or present a fire hazard. If there is a cooling fan, verify that it is working. It will not hurt to lubricate the bearings.

Check all wiring connections and retorque as needed. Loose input and output connections are a major cause of failure in this equipment.

Check input and output voltages. Input voltage should not exceed 5 percent deviation, and the three legs should be uniform.

The VFD troubleshooting protocol assumes that the electrical supply and motor have been checked out. Do not neglect the load and power transmission (shaft, belts, etc.) between motor and load. If these components are okay but the motor is not running or it is tripping out, then you have made the right choice in going into the VFD.

The following tests may be made with a multimeter.

Verify that there is no power on the dc bus. With the multimeter switched to the diode check function, place the negative lead on the positive dc bus. Place the positive lead on each of the input conductors that are connected to the rectifier diodes. They are now reverse biased, and you will read a voltage drop on each input terminal. Then place the positive lead on the negative dc bus. Put the negative lead on each of the incoming lines and see if there is a forward diode drop. From these measurements, you can determine if the diode bridge is shorted, open, or good.

If your finding is that the input rectifier diodes are not defective, you will want to check the VFD output section. Place the positive meter lead on the negative dc bus and connect the meter's negative lead to each of the three output terminals that are connected to the motor. Look for a forward-biased voltage drop. You may see that the output circuitry is either open or shorted, that is, not functioning in good semiconductor fashion. Alternately, the dc bus fuse could be blown. This may be the sole problem, or it may be indicative of a fault elsewhere.

Throughout, carefully inspect input and output devices for any sign of failure such as a burnt appearance or physical distortion.

The next step is to use the multimeter to check all capacitors for open or short. This is a likely failure mode, and is frequently apparent upon visual inspection. An oscilloscope will reveal a harmful ripple after the capacitor.

> You can make saddle, offset, or simple bends in PVC conduit (RNC) by heating it uniformly and forming the bend by eye. Do not use a propane torch; you will burn the outside before the inside is soft enough to bend. An electric blanket PVC conduit bender works very well but is expensive and prone to burn out. Various tools that use your vehicle's exhaust are on the market. Use your ingenuity and see what you can invent. I had good results holding the conduit over a charcoal grill and turning it slowly.

These are the principal diagnostic procedures for the main current path within a VFD. However, much more is involved in that operating parameters need to be programmed into the unit, and there are always the possibilities that the procedure was improperly performed or that some errant electrons caused a glitch to develop. Many technicians have problems in this area. The VFD functions in accordance with commands from the user interface, and these inputs have to find their way through the internal electronics in such a manner that the VFD will do what you want. It is a very smart machine, but despite what you may think when it is indulging in erratic behavior, it does not have a mind of its own. (We are still a few years away from machine consciousness.)

Most VFDs have onboard diagnostics as part of the user interface, and they are depicted on the alphanumeric display in the form of error codes. For the most part, the error codes are listed in the manual, and clear explanations and corrective actions are given. Most of this equipment is user-friendly and, as you expand your troubleshooting skills, you should be able to diagnose and repair these machines. Here is a suggestion: Install a telephone jack adjacent to the VFD (or use a cordless phone) and call the manufacturer. A technician will talk you through the necessary procedures.

A final word on VFD work: Lethal voltages are present throughout the drive, from power supply lines straight through to the motor windings. Even with power disconnected and locked out, the capacitors will hold a charge for a long time because they have high capacitance. You need a safe way to connect the proper load across each capacitor in order to discharge it. I suggest large insulated alligator clips, connected while wearing insulated lineman's gloves. These gloves are tested regularly by inflating them with air. A small pinhole or crack in the insulating material can expose you to hazardous electrical energy.

A further word of caution: When putting a VFD back in service, it is important not to overspeed the load, or you could send the contents of a conveyer right through a brick wall. Furthermore, you have to realize that the motor and load will be used on a daily basis, sometimes around the clock, and a malfunction down the road, built into the system and latent like a terrorist sleeper cell, could emerge at some time in the future to jeopardize the safety of workers.

Thus far, we have not had much to say about the National Electrical Code (NEC). It plays a prominent role in the professional life of electricians in areas where it has jurisdiction, although for the electronic technician the impact is somewhat smaller. One area where it becomes crucial is in the design of motor installations, especially in the workplace. Motors have to be set up correctly at the outset, and the electrical supply wiring configured in the right way, or the installation could be hazardous or fail to work at all. Setup errors can make for a very short motor life as well. High-horsepower motors, including installation costs, can run many thousands of dollars, constituting one of the major investments in an industrial location.

NEC and the Motor

NEC is vital in all of this. Nevertheless, motor design is somewhat counterintuitive, especially when it comes to overcurrent protection and ampacity calculations. As an electrician, you have to set aside temporarily what you know about "ordinary" wiring when undertaking a large motor installation. The principal guide and source of knowledge in this matter is the Code, so we will do a survey of the relevant material.

The basic fact is that motors draw vastly more electrical current during start-up than they do once the load is brought up to operating speed. If you were to provide overcurrent protection at the source of the branch circuit based on the full-load current of the motor in question, it would cut out every time you tried to start it. Therefore, it is necessary to approach the matter differently. A large motor branch circuit is protected at its source at a much higher (less sensitive) level. This overcurrent protection is valid only for branch-circuit short-circuit and ground-fault protection, not for overload at the motor.

To see how this is done, we will back up and examine in some detail NEC Article 430, Motors, Motor Circuits, and Controllers. First, we will put the venerable NEC in historical context in order to provide perspective.

If a small electric motor like in a fan will not start when energized, see if it turns hard by hand. If you give it a spin, it should coast awhile, not stop right away. If it does not turn freely, oil the front and rear bearings. Use penetrating oil at first, and then follow up with a heavier machine oil to provide lasting lubrication.

Thomas Edison and his large group of associates, in the latter part of the nineteenth century, conceived of a generation and distribution system that would bring electrical power to individual homes and businesses. The focus at the time was to provide electric lights in these buildings and outside, an improvement over existing gas lights. The concepts of fusing and overcurrent protection, even circuit breakers, had been developed decades earlier, and they were in use in telegraph systems. Therefore, the basic elements were in place. Early electricians fished wires through disused gas pipes, which became the first electrical conduit runs.

The overall system worked, and electrical power was brought into the home and workplace. However, these installations were not safe. Electrical fires and electrocutions became frequent events. Insurance companies sustained great financial loss, and they and other interested organizations developed the first National Electrical Code in 1897. A few years later, it came under the jurisdiction of the National Fire Protection Association (NFPA), which currently releases a revised edition every three years. A large number of committees and individuals work to create each new version, which contains many revisions and a large amount of new material. Electricians have to read and assimilate these new mandates and integrate them into their work on a continuous basis.

The NEC as issued by the NFPA has no legal standing on its own, but is issued so that states, municipalities, and other jurisdictions may enact it into law. Currently, the NEC is the principal electrical authority by statute throughout the United States, Colombia, Costa Rica, Mexico, Panama, Puerto Rico, Venezuela, and other locations. The Canadian Electrical Code (CEC) is substantially similar, and Europe's International Electrotechnical Commission governs electrical design and installation in that region.

The NEC is divided into an Introduction and nine chapters. Chapters 1 through 4 apply generally. Chapters 5 through 7 apply to special occupancies, special equipment, or other special conditions. Chapter 8 covers communication systems.

Chapter 4 is titled Equipment for General Use, and here we find Article 430, Motors, Motor Circuits, and Controllers. This article, one of the longest in the Code, is made up of over 50 pages covering the complex subject of motor installation. To diagnose and repair motorized equipment and the systems that support it, it is essential to be aware of Code mandates that are involved because any errors can create a hazardous situation.

Article 430 conforms to the consistent NEC template for articles. What this means is that insofar as the subject matter allows, the structure of each article is the same, with a decimal numbering system for headings and subheadings.

In this spirit, Article 430 begins with Section 430.1, Scope, and then proceeds to list definitions that pertain to the article. Next on the agenda are short discussions of part-winding motors (430.4) and Ampacity and Motor-Rating Determination (430.6). The main concept being introduced at this point is that methods for ampacity determination differ for various types of motors. Therefore, if you are planning to wire a motor to the source of power, you need to consult this section in order to determine the ampacity, circuit size, and thence the size of the supply conductors to be used.

Section 430.7, Marking on Motors and Multimotor Equipment, lists the information that manufacturers, in order to be Code compliant, must include on the nameplate. The nameplate is all-important, and the information it displays provides the starting point for designing a new motor installation.

Part II is titled Motor Circuit Conductors, and it covers the information you need to know to size a motor circuit. The basic rule is that the conductors that supply a single motor, as opposed to a group installation, are to have an ampacity of not less than 125 percent of the motor full-load current rating. How is that rating determined? It is determined by consulting Tables 430.247, 430.248, 430.249, and 430.250. These tables are found at the end of the article, and they give full-load currents for various horsepower motors. So the thing to remember is that regardless of what the nameplate may say, to find the full-load current of a motor, take the horsepower off the nameplate and then go to one of the three tables depending on the current type or number of phases. Table 430.247 is for dc motors, Table 430.248 is for single-phase ac motors, Table 430.249 is for two-phase motors, and Table 430.250 is for three-phase motors. (Two-phase systems are rarely used and are mostly obsolete.)

Section 430.6(2), Nameplate Values, states that separate motor overload protection is to be based on the motor nameplate current rating.

This is the first mention of separate overload protection, and it is the key concept in wiring motors because it provides the procedure for protecting motors from overcurrent while at the same time allowing for the very high inrush starting current that motors require.

An interesting and valuable table is 430.7(B), Locked-Rotor Indicating Code Letters. This is among the information required to be displayed on the motor nameplate, and it consists of a single letter, A through V. (I is not included because it could be confused with the number 1.) Each letter corresponds to a certain number of kilovolt-amperes per horsepower drawn by a motor with a locked rotor. This becomes clear once a couple of points are clarified.

Kilovolt-amperes are similar to kilowatts. One volt-ampere is equal to 1 watt in a dc circuit, but the former takes into consideration the effect of frequency. At 60 Hz, the effect is not too great, but volt-ampere terminology is more technically correct.

The values given in the table are per horsepower, so this means they are consistent for any size motor—large or small. Moreover, the figures are applicable to a motor with a locked rotor, that is, a rotor that is mechanically fixed so that it cannot turn, or encumbered with a very heavy load for the size of the motor so that it will fail to start. Of course, the locked-rotor current will be quite large, especially for a big motor. Going down the code letter alphabet, the locked-rotor current becomes higher. Code Letter A motors have a maximum locked-rotor current of 3.14 kilovolt-amperes per horsepower, while Code Letter V motors have a minimum locked-rotor current of 22.4 kilovolt-amperes per horsepower. This represents a very large range. A motor with the lower locked-rotor current is a higher impedance motor. Generally, motors should not be run with the rotor locked or they will quickly burn out. However, with sufficiently high impedance and low supply voltage, a motor can operate indefinitely in a locked-rotor state without overheating. Such a motor is called a "torque motor."

Often we will see more than one motor on a single circuit. These are called "group installations." Part II includes information on wiring them. It states that conductors supplying several motors, or a motor and other loads are to have an ampacity not less than the sum of each of the following:

1. 125 percent of the full-load current rating of the highest-rated motor.

2. Sum of the full-load current rating of all the other motors.

3. 100 percent of the noncontinuous non-motor load.

4. 125 percent of the continuous non-motor load.

Part III, Motor and Branch-Circuit Overload Protection, describes the overload protection, which may take the form of a fuse or circuit breaker. Its purpose is to protect the motor from failure to start due to excessive or defective load, single phasing of a three-phase motor, or other adverse conditions including problems inside the motor. This protection is in addition to and separate from the branch-circuit short-circuit and ground-fault protection at the source of the branch circuit.

To select the overcurrent protection device, we use the motor nameplate full-load current rating, not the value from the tables at the end of the article, which were used to select the branch-circuit short-circuit and ground-fault protection device.

The device is to be selected to trip at no more than the following percentage of the motor nameplate full-load current rating:

Motors with a marked service factor 1.15 or greater: 125 percent.

Motors with a marked temperature rise of 40°C or less: 125 percent.

All other motors: 115 percent.

Service factor is one of the parameters marked on the motor nameplate. Motors with a larger service factor are more conservatively rated. If you are confronted with an installation where a motor consistently runs hot or trips out, you may want to go to a motor having a higher service factor rather than a motor of higher horsepower. Service factor has to do with the heat tolerance of the internal motor insulation.

It is further provided in Part II that where the sensing element or setting or sizing of the overload device is not sufficient to start the motor or to carry the load, higher size sensing elements or incremental settings or sizings are permitted to be used, provided the trip current of the overload device does not exceed the following percentage of motor nameplate full-load current rating:

Motors with a marked service factor 1.15 or greater: 140 percent.

Motors with a marked temperature rise 40°C or less: 140 percent.

All other motors: 130 percent.

The most important provision having to do with motor overcurrent protection (because it allows motors to actually start) follows.

If not shunted during the starting period of the motor, the overload device is to have sufficient time delay to permit the motor to start and accelerate its load.

An Informational Note states that a class 20 or class 30 overload relay will provide a longer motor acceleration time than a class 10 or class 20 relay, respectively. Use of a higher-class overload relay may preclude the need for selection of a higher trip current.

Table 430.37, Overload Units, specifies the number and location of overload units, such as trip coils or relays.

The numbers and locations vary depending on the current type (ac or dc) and the phase of the motor. What is surprising to many apprentice electricians is that the overload relay may be in just one hot conductor, for example, in a 240-volt single-phase supply. This is because the overload is not intended to perform the disconnect function. However, in a three-phase system there must be an overload in each of the three legs so that the motor will not attempt to run and be damaged if one leg is lost.

Part IV, Motor Branch-Circuit Short-Circuit and Ground-Fault Protection, specifies devices that will protect the entire circuit, including controls, through to the motor. It protects them against short circuits and ground faults at a high level, but does not protect the motor against overloads, which is done at a lower level with time delay built into the overload protective device.

Table 430.52 gives the maximum rating or setting of branch-circuit short-circuit and ground-fault protective devices for seven types of motors, using four types of protective devices. These ratings are given as percentages of full-load current. You will notice that they are quite high. For example, a single-phase motor can be protected at the branch-circuit level at 175 percent of the full-load current using a nontime delay fuse, 250 percent with an inverse-time breaker, 300 percent with a nontime delay fuse, and 800 percent with an instantaneous trip breaker!

Additionally, where these values do not correspond to standard sizes, it is permitted to go to the next higher rating.

On the subject of several motors or loads on one branch circuit, Article 430.53 lays out rules for single-motor taps. NEC tap rules are sometimes difficult for electricians, both on licensing exam and actual design/installation levels, because these rules do not appear in any one Code location. They are spread throughout this heavy volume, which means you have to know where to look. The motor tap rules provide that for motor group installations, the conductors for any tap supplying a single motor are not required to have an individual short-circuit and ground-fault protective device, provided they comply with one of the following:

1. No conductor to the motor is to have an ampacity less than that of the branch-circuit conductors.

2. No conductor to the motor is to have an ampacity less than 1/3 that of the branch-circuit conductors, the conductors to the motor being not more than 25 ft long and being protected from physical damage by being enclosed in raceway.

3. Conductors from the branch-circuit short-circuit and ground-fault protective device to a listed manual motor controller additionally marked "Suitable for Tap Conductor Protection in Group Installations" or to a branch-circuit protective device are permitted to have an ampacity not less than 1/10 the rating or setting of the branch-circuit and ground-fault protective device. The conductors from the branch-circuit and ground-fault protective device to the controller are to (a) be suitably protected from physical damage and enclosed either by an enclosed controller or by a raceway and be not more than 10 ft long, or (b) have an ampacity not less than the branch-circuit conductors.

You will notice, in interpreting tap rules, that there is a range of permitted reduced sizes available from which to choose. In addition, to compensate there is a range of techniques for required isolation from damage and maximum lengths permitted. For example, with the 1/3 reduction, you can run a 25-ft tap, while with the 1/10 reduction you can go only 10 ft. All Code tap rules are similar, although they differ in the details.

Part VI, Motor Control Circuits, is especially applicable to the troubleshooting endeavor. The controller enhances motor function and generally works quite well, but there are times when, rightly or wrongly, it will prevent the main supply current from getting through to the motor. Controllers are easy to diagnose and service because, aside from the main contacts, the amount of current and wire sizes are modest and the controller should be in a good accessible location.

First, we will look at NEC mandates for motor control circuits, and later we will discuss troubleshooting and repair techniques for motor controllers and associated wiring.

Generally, a motor control circuit is tapped from the load side of a motor branch-circuit short-circuit and ground-fault protective device. It functions to control the motor connected to that branch circuit and, of course, the primary motor overcurrent device is much too large to protect the controller, a relatively light load with small conductors, so the conductor must be protected separately. A tapped control circuit is not to be considered a branch circuit, so it may be protected by either a supplementary or a branch-circuit overcurrent device.

Table 430.72(B), Maximum Rating of Overcurrent Protective Device in Amperes, gives these ratings as a function of the control circuit conductor size. The general Code rule is that the minimum conductor size for any circuit is 14 AWG, but this table states that control circuit wires can be 16 AWG or 18 AWG, as long as they are protected by overcurrent devices as shown.

Where a motor control circuit transformer is provided, the transformer is to be protected as follows:

1. Where the transformer supplies a Class 1 power-limited circuit, Class 2 or Class 3 remote-control circuit complying with the requirements of Article 725, protection must be in accordance with that article.

This raises the complex subject of Class 1, Class 2, and Class 3 remote control, signaling, and power-limited circuits. Requirements are found in NEC Article 725, and it is recommended that individuals interested in troubleshooting and repairing commercial electrical equipment study this topic in detail. (It is discussed in my previous McGraw-Hill book, *2011 National Electrical Code Chapter By Chapter*.)

The Relevance of Article 725

Article 725 is frequently misconstrued as dealing with "low-voltage wiring" but this is a misnomer. The subject of Article 725 is remote control, signaling, and power-limited circuits; the voltages (in Class 1 circuits) can run as high as 600 volts and there is not always a power limitation. In some cases, however, there are power and voltage limitations. Accordingly, Article 725 permits a relaxation of some specific Code requirements in some applications, while in other applications more rigid forms of protection are mandated. Motor control circuits may be permitted to conform to any one of these classes, so do not be too surprised if you see unusually small (less than 14 AWG) conductors with splices made outside of enclosures.

1. Compliance with Article 450, Transformers . . . (Including Secondary Ties).

2. Less Than 50 Volt-Amperes—Control circuit transformers rated less than 50 volt-amperes and that are not an integral part of the motor controller and located within the motor controller enclosure are permitted to be protected by transformer primary overcurrent devices, impedance-limiting means, or other protective means.

3. Primary Less Than 2 Amperes—Where the control circuit transformer rated primary current is less than 2 amperes, an overcurrent device rated or set at not more than 500 percent of the rated primary current is permitted in the primary circuit.

4. Other Means—Protection is permitted to be provided by other approved means. "Approved" is Code terminology. It means allowed by the local electrical inspector or other authority having jurisdiction.

Section 430.74, Electrical Arrangement of Control Circuits, brings up a significant safety issue. It states that if one conductor of the motor control circuit is grounded, the motor control circuit is to be arranged so that a ground fault in the control circuit remote from the motor controller will not start the motor nor will it bypass manual or automatic safety shutdown devices.

Part VII, Motor Controllers, lays out requirements for controllers, as opposed to motor control circuits, covered previously. The overriding principle, stated at the outset, is that all motors that are powered by a premises electrical system must have controllers. However, for a stationary motor of 1/8 horsepower or less that is normally left running and is constructed so that it cannot be damaged by overload or failure to start, such as clock motors and the like, the branch-circuit disconnecting means, usually a circuit breaker, will serve as the controller.

Similarly, for a portable motor rated at 1/3 horsepower or less, the attachment plug and receptacle or cord connector will serve as the controller.

Larger motors require more advanced controllers, and the principle is that any motor controller must be capable of starting and stopping the motor it controls, and interrupting the locked-rotor current of that motor.

A branch-circuit inverse time circuit breaker rated in amperes is permitted to serve as a controller for any motor. In addition, a molded-case switch rated in amperes may serve as controller for any motor. A molded-case switch looks like a circuit breaker and it fits in a circuit breaker enclosure, receiving its power from the busbar and having screw terminals for outputs. It does not necessarily include overcurrent protection. It may serve as a motor disconnecting means, and must be sized at 115 percent of the motor full-load current.

Other devices are also permitted as motor controllers, notably air-break ("ordinary") switches, inverse-time circuit breakers, manually or power operated, and oil switches.

It is further provided that every motor must have an individual disconnecting means, although exceptions allow a single disconnecting means to serve a group installation under certain conditions, such as when a group of motors is in a single room within sight of the disconnecting means.

Part X, Adjustable-Speed Drive Systems, uses this generic term to cover the VFDs that we discussed earlier in Chap. 2.

The point is made right away that the installation requirements of the previous nine parts of Article 430 are applicable to Part X except where modified or supplemented therein.

Section 430.122, Conductors—Minimum Size and Ampacity, states that the electrical supply conductors for the power conversion equipment that is part of the adjustable drive system are to have an ampacity not less than 125 percent of the rated input current to the power conversion equipment. This means that you go by the adjustable drive system nameplate, not the motor nameplate, and add an additional 25 percent for good measure.

Some adjustable drive systems incorporate a bypass device, where power from the electrical supply can be shunted directly to the motor. The supply conductors for these systems are sized based on 125 percent of the rated input current to the adjustable drive system or 125 percent of the motor full-load current rating, whichever is higher.

Section 430.124, Overload Protection, states that overload protection for the motor is required. Some adjustable-speed drive systems include the overload protection and, if it is marked to so indicate, additional overload for the motor is not required. However, if a bypass device is installed to allow the motor to operate at full load speed, then overload protection must be included in the bypass circuit.

Section 430.126, Motor Over-Temperature Protection, makes note of the fact that the relationship between motor current and motor temperature changes when a motor is operated by an adjustable-speed drive. There can be a problem when an adjustable-speed drive causes a motor to run slower than its rated speed, especially when there is a fan attached to the motor shaft. There will be less cooling, and the motor can overheat. Protection against motor heating is to be provided by one of the following:

1. Motor thermal protector
2. Adjustable-speed drive system with load- and speed-sensitive overload protection and thermal-memory retention upon shutdown or power loss
3. Over-temperature protection relay utilizing thermal sensors embedded in the motor
4. Thermal sensor embedded in the motor whose communications are received and acted upon by an adjustable-speed drive system

Additional adjustable-speed drive provisions apply to equipment operating at over 600 volts, protection of live parts, and grounding.

The 2011 NEC section that covers adjustable-speed drives is brief because the Code does not have jurisdiction over the internal workings of factory-made electrical equipment. It is up to the authority having jurisdiction to approve or not approve a given piece of electrical equipment. Since the local inspector lacks the test equipment and knowledge to evaluate each factory-made unit in an installation, he or she will rely on listings by a testing organization such as Underwriters Laboratories, which has extensive testing facilities and resources to evaluate products.

This part of the Code, therefore, applies primarily to the branch circuit wiring that supplies power to the adjustable-speed drive, and to the conductors that run from the output of the drive to the motor.

> The National Electrical Code says a run of conduit has to be put in as a complete system, terminated at both ends, before installing conductors. You can put in a pull rope as you hook up the conduit lengths, but not conductors. Sometimes the easiest way is to use a shop vac to pull a piston (so-called "mouse"). This is a foam rubber cylinder sized to fit the pipe, with string attached. If a piston is not available, make your own by carving it out of foam or similar material.

The foregoing has been a brief survey of Article 430, Motors, Motor Circuits, and Controllers. There are many provisions that have not been mentioned here. If your troubleshooting and repair involve any alteration of the original installation, it is suggested that you carefully review Article 430 and, indeed, the entire Code to ensure that all elements of your work are in compliance.

Concerning troubleshooting a motor controller and repairing it, these devices are not overly complex for many ordinary applications. You should be able to look at the controller and understand how it works without need of a manual or schematic. However, for a more complex controller, full documentation is required. The controller may be designed to start, stop, reverse rotation, change speed or torque, or any number of other motor parameters. A timer may be included to control the duration of sequential actions, and some elaborate controllers sense motor overload and temperature.

Types of Motor Controllers

Controllers may be human operated from one or more remote stations, or they may automatically receive instructions relating to position and the state of other sensors that are part of the driven equipment. As a generic example, we can conceive of a motor controller that has two electrical circuits—a control circuit powered by a low-voltage source such as a 24-volt transformer and a power circuit connected to the motor. The control signal will activate a magnet that will close three sets of contacts to power a three-phase motor. Schematics often do not show the power circuit in the interest of simplicity.

The troubleshooting procedure, after looking over the controller for visible signs of damage, is to check input and output voltages of the power circuit with the contacts open and closed. If the power circuit appears functional but the contacts will not pull in, the magnet could be at fault or it is not being energized. See if there is voltage at the transformer secondary. (The transformer is probably located some distance away from the controller, between the load center and the actuator.) If there is no voltage at the secondary, check the primary. If you find the transformer is dead, which is often the case, the problem is solved. If the transformer is good, just out of curiosity compare the measured secondary voltage to the marked secondary voltage on the transformer. A few volts difference is not critical, but if there is a large discrepancy, there are shorted windings, and this can cause the controller to become erratic. The most common transformer fault, however, is zero volts at the secondary—an open primary or secondary. Of course, the whole thing could be a tripped breaker supplying the primary, perhaps indicating a shorted power supply line. In addition, there may be a low-ampere plug fuse associated with the transformer. Ohm readings, with the transformer taken out of circuit, are useful as well.

If the transformer is good, now that you know the secondary voltage put your meter on the controller input with the actuator properly set. If you do not read voltage, unhook one wire and try an ohm reading. It is possible that the controller input is shorted, pulling the voltage down to zero, the transformer and line having sufficient impedance so that the current level would not become too high.

In an industrial setting, with vibration and corrosion possible factors, control line faults are common. If you disconnect the line at both ends and temporarily tie the wires together at one end, you can do an ohmmeter check.

If there is voltage at the control input but the power contacts do not close, there can be an internal wiring fault or defective coil.

Returning to the power circuit, a frequent fault is bad relay contacts. A prelude to this is an audible 60-Hz hum. You can sometimes renew the contacts by pressing them gently on a thin file and rubbing the file back and forth. Afterward, a piece of paper has just the right abrasive property to burnish the contacts. Avoid using sandpaper because it can leave particles embedded in the metal, which will later make hot spots.

If wear is pronounced, the contacts can be replaced. Contact kits are readily available.

We have been talking about magnetic mechanical relays. Many of them are still in service. The overloads are often "heaters," which come in various ampere ratings and are easily replaced.

Today many controllers are solid-state and they are quite reliable, although faults may occur.

If the motor is cutting out periodically, check the heater value or solid-state overload setting against the motor full-load current (with permitted Code adjustment) to find out if the motor is protected at the correct level. Remember that a motor will draw more current and run a little hotter as it ages, and some of this is acceptable.

In industrial facilities, motors are a major part of the environment, and maintenance electricians spend a good part of their time working on them. You need to become a motor expert. Examine, visually and with test equipment, good running motors and keep dated records of the results. Read the documentation and, if warranted, make sure that good replacement motors are stocked. Where used motors are available for replacement purposes, clean and test run them, and label them. Check the bearings and if there is significant play, replace them.

Residential Wiring

T his book is mostly about commercial troubleshooting and repair, but we include this chapter on residential wiring in order to provide perspective. For the electrician, residential wiring is usually the gateway into the field. Most technical-oriented people are familiar with the infrastructures of their own homes, where they do routine repairs and maintenance as needed, including working on heat systems, air-conditioning, garage door openers, etc. An apprentice or beginning electrician is likely to work for an electrical contractor who specializes in residential work. Beginners are given menial but important tasks like moving materials and cleaning, and those with aptitude and ambition quickly graduate to actual electrical work.

Residential electrical installation differs markedly from commercial and industrial work. Often it involves new construction on a large scale, as when extensive subdivisions are built. Every house may be essentially the same electrically, differing only in the location of the service due to differences in lot orientation. Much of the work involves repetitive tasks like drilling holes in studs and running wire between outlets. Speed and efficiency are of the essence, and supervisory personnel will urge the workers on to ever-higher levels of productivity. The commercial or industrial maintenance electrician, in contrast, must have a greater amount of knowledge and expertise to design, install, repair, and maintain the vast amount of equipment in use, but strangely the pay is often lower, and the work pace a little slower and more deliberate.

In residential work, the somewhat more complex work of alarm, communication, and data wiring is often but not always done by specialized subcontractors, so that the construction electrician is left with the simpler, but no less exacting, tasks of service and branch-circuit design and installation, and putting in light fixtures and fixed appliances.

In residential wiring, there are enormous moral and legal concerns. Safety for coworkers and end users must always be before the electrician. Think of the young child sleeping in a third-floor bedroom of a building you have wired.

To mount a fluorescent strip fixture on a sheet-rocked ceiling, it is necessary to screw through the drywall into framing. Do not worry if the holes in the fixture do not coincide. Use your cordless to screw right through the sheet metal without regard to the predrilled holes.

It is this precise issue that National Electrical Code (NEC) addresses. Protection from fire and electrical shock is the focus of the thousands of statements and table entries that comprise this work. It is fundamental to all electrical construction, and has great importance in the residential setting.

Like any wiring, having an informed approach begins with the design. The first step in residential work is to size out the electrical installation, based of course on the size of the house and its anticipated usage. The complete procedure for this process is found in NEC Chap. 2. The electrical load depends upon the number of square feet of the house, to which is added the load of each appliance, with exceptions and permitted derating factors. For example, only the larger of two loads that are not likely to be run concurrently must be included in the calculation of service and feeder conductor sizes. The classic example of this is heating and air-conditioning.

However, since our main interest is troubleshooting and repair, we will not undertake a close study of this important topic.

Problem Solving

It is a frequent occurrence for the electrician to be called in at some point after the new owners have taken possession of the premises, even decades later for older houses, to fix problems that have surfaced. Often the homeowner has attempted to do the work, but has been unable to carry it to completion. The same troubleshooting techniques are applicable in residential wiring systems as in electrically driven heavy machinery.

First, interview the owner/occupant. Find out whether the problem in the wiring always existed, suddenly appeared, or gradually emerged. If it existed right from the start, it probably resulted from a fault in the installation that was not serious enough to be noticed right away. An example of this is the battery-operated smoke alarm with backup ac that was initially bugged off the power supply to a light fixture, so that the ac mode works only when the switch is turned on.

If the problem emerged gradually, for example flickering lights, at first barely perceptible but eventually quite prominent, then the fault is time-dependent, and it is likely the result of a gradual increase in impedance at a single point in the wiring. This may be the result of corrosion, and it is a simple matter to clean or remake the connection, but the difficulty is often locating that point. (Each of these faults will be discussed shortly.)

A whole other category involves ground-fault circuit-interrupter nuisance tripping. This is a subset of overcurrent device tripping out in general, although the causes are quite different. Another electrical fault involves erratic or immediate tripping out of arc-fault breakers, a problem made more difficult because the fault is often concealed beneath wall or ceiling finish. Just about the only residential circuitry that is at all complex (aside from alarm and low-voltage work) is that involving three-way and four-way switching. Homeowners who successfully install their own wiring for an addition or garage sometimes become hopelessly confused when wiring these devices, and they call in an electrician to make sense of things.

In addition, at times, the electrician is asked to solve a real or imagined situation of chronic high electrical usage as reflected in the utility bill, and there are instances where this is caused by poor premises grounding. Another culprit could be an individual inefficiently functioning appliance.

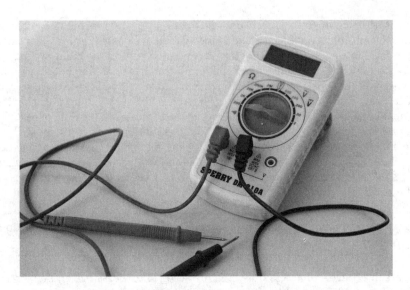

FIGURE 3-1   Inexpensive multimeter.

A principal troubleshooting technique is half splitting, which was discussed earlier. This method is applicable in residential work, and will quickly take you to the fault. Several tools are useful in residential troubleshooting. The clamp-on ammeter and the loop impedance meter are a great help, as is the circuit analyzer and multimeter, especially in the ohms mode (see Fig. 3-1).

Mentioned previously were some common residential wiring faults. Here they are discussed in more detail with the best troubleshooting techniques.

A Hazardous Situation

Flickering lights are a rather serious problem, and are often a prelude to a catastrophic fire. The homeowner notices a slight intermittent flicker that at first appeared to be an isolated incident. Gradually, it becomes more frequent and intense, so an electrician is consulted. It is likely that the homeowner has heard, or figured out, that the problem is serious, due to a poor connection that is arcing and could generate enough heat to ignite nearby combustible material. On the other hand, it could be due to a fault in the utility supply, for example, a secondary connection at the transformer in which case the utility should be notified so that they can make the repair.

Since the fault is intermittent, it may not be happening when you visit the site, but most likely you can locate it anyway. To begin, find out the extent of the impacted area. This can be done by observation, if the intermittent is active, or by questioning the occupants. If the flickering is confined to a single cord-and-plug-connected fixture, the homeowner would probably have repaired or discarded it and the electrician would not have been called in. However, it could be confined to a hardwired ceiling fixture controlled by a wall switch. Before going after a ladder, wiggle the switch from side to side while it is in the ON position. If the flickering responds, you have a bad switch or bad connection at the switch.

If there is no reaction at the switch, and the light fixture appears not to contain the fault, and if your power outlets on that specific branch circuit are unaffected, you have to assume that there is fault in a part of the wiring that is concealed behind the building finish—drywall, wood paneling, or other wall or ceiling finish. These are rare, but they do happen. Has there been recent construction activity whereby an errant nail could have compromised a current-carrying conductor? Alternatively, perhaps a wire was damaged during the initial building, making a local hot spot that finally got worse because of heating at that point.

It is a basic Code principle that spliced conductors, within enclosures, must always be accessible, that is, the cover has to be capable of being removed to allow access to the interior of the enclosure. Junction boxes buried behind wall, ceiling, or other finish were at one time called "blind boxes," but this term is no longer used out of respect for visually impaired individuals (see Fig. 3-2).

> Wire nuts in questionable environments should be installed with the openings down so that any moisture will drain.

There are three levels of accessibility: readily accessible, accessible, and not accessible. NEC defines accessible as capable of being removed or exposed without damaging the building structure or finish or not permanently closed in by the structure or finish of the building. Readily accessible means capable of being reached quickly for operation, renewal,

FIGURE 3-2
4×4 box with Romex cable.

or inspection without requiring those to whom ready access is requisite to climb over or remove obstacles or to resort to portable ladders and so forth. The classic example of a location that is accessible but not readily accessible is above a suspended ceiling. Junction boxes are permitted here, and it is an easy task to pop out a panel and peer around to find any junction boxes and service the splices within them.

Splices or terminations that are accessible often cause flickering lights. If there are junction boxes that might contain bad connections, the procedure is to wiggle, probe, and tighten the wire nuts, all the while watching or having a helper watch the load for any reaction.

Sometimes wire nut splices go bad because they were not tightened sufficiently to begin with. Alternatively, especially if there are more than two conductors, they can eventually start arcing if they were stuffed into an overcrowded enclosure—one reason we have box-fill rules. In these instances, the splice can be remade and tightened. However, if there is any sign of oxidation or if there has been flooding or moisture from condensation or a leak, new wire nuts and new wire terminations will be needed. In this connection, it is always a good practice, on the initial installation, to place wire nuts with the opening pointing down, so any condensation or other moisture will drain.

While we are on the subject, some workers wrap electrical tape around their wire nuts. In my view, it is a bad practice and will probably void the testing laboratory listing. In the first place, those wings and ridges are cooling fins. Tape will hold the heat and cause temperature rise at the splice, which is exactly what you do not want. In addition, the tape adds extra box fill, and that is not good either.

What about twisting the wires prior to putting on the wire nuts? Some electricians are twisters and some are not. I am against that idea also, except for stranded wire or where there are more than three wires where twisting is needed to keep them together. Twisting kinks and hardens the wire so it is difficult to remake the connection if that becomes necessary later. One of the good things about wire nuts is that they can be removed easily for testing and alterations. To make a wire nut splice, just lay the pieces of solid copper alongside each other with the ends perfectly aligned, and let the wire nuts do the work. If necessary, hold the wires with smooth-jawed needle-nose pliers so the wires remain where they belong. When stripping insulation, take off the right amount so that no copper shows when the wire nut has been twisted on and so that there is no insulation inside the metal cone to impede the conduction.

If there is no fault in a junction box, fixture, or switch in the branch circuit, go back to the entrance panel or load center and find the breaker that powers the circuit. Wiggle it and see if the flickering reacts. If so, the fault is invariably where the breaker connects to the bus bar. Arcing and corrosion resulting from the heat conspire to pit and burn the breaker contacts and bus bar. At first, it will be a onetime event, but once the process starts, it invariably worsens, and the heat increases, especially if the load is heavy. Causes of this unfortunate occurrence are moisture in the box, repeated short-circuiting of the breaker, vibration, and running the box with no cover. (The cover helps hold the breakers firmly in place.)

A bad breaker should be discarded. It might be possible to clean up the bus bar (with main breaker off) using Scotchpad, but if the pitting is significant, the bad position in the box should not be used. The disadvantage in merely shifting positions, however, is that at some time in the future someone could insert a breaker at the affected position, setting the stage for another event.

When bus bars are damaged in this way, most electricians discard the entire box and install a new one. This is not necessary. Bus bar replacement kits are available for a small fraction of the price of the box. Be sure to get the model and serial number of the old box because there are variations within the same make.

If roughly half the circuits in the building are flickering, it is time to look closely at the main breaker. This is similar to the branch-circuit problem, but far more dangerous because the load is heavier so that there are more electrons trying to squeeze through the high-impedance bottleneck. Enough heat can be generated to pass through the back of the box and ignite the wall material behind it. Moreover, an outage that takes place before repairs can be made will put the entire building in darkness or without heat (see Fig. 3-3).

Wiggle the main breaker from side to side to see if the flickering reacts. If the main is bad, there will be a loud frying and popping sound, and light will be seen coming from behind the breaker. If this is happening, it is imperative that the box be de-energized immediately and be kept out of use until the repair is made because there is a severe fire hazard.

In a single-phase main, two hot conductors carry power from the utility. The conductors are usually aluminum rather than copper. Aluminum compounds the problem. (Aluminum is much less expensive; otherwise, copper would always be used.)

For a given ampacity, aluminum is always required to be larger in accordance with NEC tables. For example, a 200-ampere residential service calls for 4/0 aluminum or 2/0 copper. Even in the larger sizes, aluminum is problematic. For very short runs where cost is not a concern, it is better to use copper. For long runs in large sizes, the cost of copper is prohibitive.

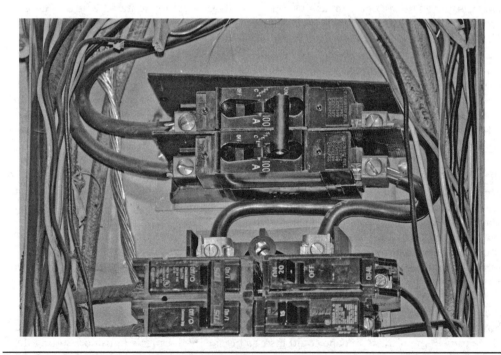

Figure 3-3   Circuit breaker box with main disconnect.

The problem with aluminum aside from the ampacity, which is corrected by going to the larger wire, is twofold. For one thing, it is less dimensionally stable; this means that if you tighten a termination, the aluminum will seem to flow. After a couple of years, the lug will appear to have loosened, setting the stage for arcing. Moreover, aluminum oxidizes, especially if there is any heat. The oxide is not a good conductor, so there is impedance and more heat.

One remedy is always to apply an oxide inhibitor such as Noalux upon initial installation, preceded by wire brushing.

It is possible for the entire building to be subject to flickering, that is, for both legs to be affected. This happens when the neutral is arcing. That is less frequent, however, because in the neutral there is less current. The neutral carries the difference between the two legs' currents, since they flow in opposite directions and cancel out each other.

If work is to be done on the main, it is necessary to shut off the power at some point upstream. This is usually accomplished by removing the meter from the meter socket unless there is a separate main disconnect. You have to call the utility and get permission to pull the meter because this involves breaking their seal. (If a meter reader finds a broken seal, an investigation will be initiated to see if someone is stealing unmetered power.) The utility should give you a plastic shield, which is used to cover the live parts inside the meter enclosure. The uncovered meter socket, with live upper lugs, should never be left untended. With the meter removed, there will be no power on the main, and it will be safe to replace the entrance panel and remake the connections.

In some instances, the entrance panel including main breaker will not be at fault. A loose connection in the meter socket or weatherhead is rare, but sometimes occurs. There can be a fault at the transformer secondary. Utilities define the customer's point of connection, and this varies in different areas, so upstream from the main breaker it is necessary to check with them to see who is responsible for the repair.

Sometimes it is difficult to find the location of the fault when there is an erratic leg. One clue is to see if the flickering occurs when the wind is blowing, which would indicate a service-drop conductor (Triplex cable) rubbing on a tree limb. Another good method for narrowing down the location of a fault when the flickering is active is to connect a voltmeter to the input of the main with the main breaker off and see if you get fluctuations. Using common troubleshooting techniques, you can find the bad connection and correct a very hazardous condition.

> Before commencing work on a light fixture directly over a sink, close the drain. If a screw or small part is dropped, it will not be lost.

We mentioned that three-way and four-way switching could be a challenge for homeowners and beginning electricians. Many times the mistake is in the neutral. The purpose of three-way switches is to permit operation of a load, usually a ceiling fixture, from two separate locations. Examples are a room with two entries at opposite ends, a stairway with one three-way switch at the top and one at the bottom, and a detached garage with the lighting capable of being controlled in the garage or in the house. Any number of additional control locations is possible by adding four-way switches between the pair of three-way switches.

In wiring three-way switches, several configurations are possible, depending upon the layout of the room. The most basic setup is when power, from an entrance panel or load center, is delivered to a wall box enclosing the first three-way switch. Cable from a different knockout in this box is run to a wall box housing the second three-way switch. Cable from a second knockout in this box runs to the light fixture.

A Thought Experiment

The best way to think about the two three-way switches is as a "black box" that functions as a regular single-pole switch. A black (hot) conductor feeds one terminal, the input, of this imaginary composite switch and a black conductor (switched hot) is connected to the output of the switch and powers the light fixture when the switch is in the ON position. White, the neutral return from the light fixture, runs straight through from the light fixture back to the power source. It is never connected to a switch, although it is spliced with wire nuts in both boxes.

What makes a pair of these switches different is that the two parts of the black box are in separate locations, so the two three-way switches have to be wired together. 2011 NEC ampacity tables provide for small loads, up to and including 15 amperes, to be supplied by 14 AWG conductors. However, many electricians feel that this wire is too small even for these residential branch circuits, and they use 12 AWG copper. Alternatively, maybe it is that they want to have a more consolidated inventory. For whatever combination of reasons, 14 AWG is not always used where permitted. Despite the practice, 14-3 AWG is used usually to wire from one three-way switch to the next. (Three conductors are required, so that what is needed is 14-3 AWG with ground. In this numbering convention, the equipment grounding conductor is not counted, so 14-3 AWG actually has four wires, and 12-2 has three. Throughout these discussions, it is assumed that the equipment grounding wire will be connected at each device.)

Three conductors are required between the two three-way switch boxes. The red and black are two alternate ungrounded current paths, and the white is either the grounded neutral or the switch-loop return, in which case it is reidentified black, as explained next (see Fig. 3-4).

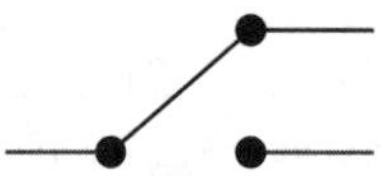

Figure 3-4
Three-way switch schematic.

The red and black of the 14-3 AWG are called "travelers," but I call them "politicians" in order to inject a little humor into an otherwise dry topic.

The position of the first three-way switch (up or down) determines which of the politicians is energized, and the position of the second three-way switch determines whether the load is energized. Thus, the switch handles in the ON position will alternate between being up and down, so they are not labeled ON or OFF.

That is all there is to it, except knowing how to terminate the wires at the three-way switches. A three-way switch always has three terminals (not counting the equipment grounding terminal) and the screws have a brass or dark finish, not silver. What can be confusing is that one of these is labeled "common." We usually think of a common terminal as connected to one input and one output wire, these being joined so that two wires can do the job of three. For three-way switches, this is not the case. The common terminal of the first three-way switch is connected to the incoming black (hot) wire from the entrance panel or load center. The common terminal of the second three-way switch is connected to the outgoing (black) wire that goes to the load. You will notice that this common terminal is located at one end of the switch body and the two terminals that connect to the red and black politicians are at the other end. If you come across a three-way switch that is constructed differently, you can identify the terminals by ringing out the switch with an ohmmeter.

As for the other configurations, it is sometimes desirable to bring the branch-circuit power directly to the light fixture, and wire the two three-way switches as if they were one single-pole switch loop. For this procedure, 12-2 cable is run from the power source to the fixture, where the neutral (white) connects to the appropriate terminal. Remember that the neutral from the power source always connects to the neutral terminal of the load, without any involvement with the switching, except that in the first configuration, it passes through the switch boxes.

In the segment of 12-2 cable that goes from the fixture to the first switch, the white is not a neutral. It is the return conductor of the switch loop of which the two three-way switches are a part. The white conductor is to be reidentified as black at both ends. This marking may be done with paint or black tape. It must completely encircle the conductor. Some electricians make three rings of tape, but just one is compliant. Colors other than black are permitted for the switch loop, but not white or green. All of this assumes that the residential building is being wired in Romex, as is usually the case. If it is being wired in electrical metallic tubing (EMT) or other raceway, the correct colors will be pulled in the first place, and reidentification will not be necessary.

A third configuration is when line power is run to the second three-way switch. In this situation, 12-2 is run to the fixture, the white being the neutral and the black being a switched hot. Here are a few points to keep in mind:

- The fixture always requires a white (neutral) wired directly back to the source and never switched, although in cases where power is not run initially to the fixture, the neutral will be spliced in one or more of the switch boxes.

- When power is initially run to the first switch, the whites are all neutrals.

- When power is initially run to the light fixture, the whites are all switch-loop returns, and are reidentified black.

- When power is fed to the three-way switch that is closer to the light fixture, the white conductor that is part of the 12-2 cable is neutral. The white conductor that is part of the 14-3 cable is a switch-loop return, and is reidentified black.

- The red and black that are part of the 14-3 are called travelers or politicians, and they are connected to the two terminals that are opposite one another at one end of the three-way switch bodies. It does not matter which is red and which is black, but they are never connected anywhere else. To the single terminal at one end of the switch body (labeled "common"), you connect only a black (hot) wire of the 12-2, or a white switch-loop return of the 12-3, which is reidentified black.

The pair of three-way switches described above permits the user to operate a load from two locations. Any number of additional locations can be added simply by installing four-way switches between the three-way switches. The four-way switches are simpler to wire than three-way switches. They have two input terminals at one end of the switch body and two output terminals at the other end. The four-way switches go with the politicians—a red and a black at each end. It does not matter which is red and which is black, nor does it matter which end of the switch is the input and which is the output. A white conductor is never connected to the four-way switch.

With the above principles in mind, you will never have a problem figuring out how to wire these switches, and you will be able to help the unfortunate homeowner who has become hopelessly confused or the beginning electrician. Most residences have two or three pairs of three-way switches. Four-way switches are rarely used, but when called for they are a worthwhile enhancement.

Ground-fault circuit interrupters (GFCIs) are used extensively in residential and other installations. Since their introduction in the 1970s, each Code cycle has mandated additional locations where their use is required, and with good reason. These simple devices are great lifesavers as they provide excellent protection from line-to-ground electric shock (see Fig. 3-5).

A GFCI may take any of several forms. One type is a GFCI circuit breaker. It occupies a breaker position in an entrance panel or load center and provides conventional branch-circuit over-current protection as well as GFCI protection. Like all GFCIs, it protects all downstream wiring, but has no effect on the upstream wiring. This device inputs its power from the busbars of the box, and has a terminal screw for the output. Additionally, a white pigtail is connected to the neutral bar. Like all GFCIs, there is a test button to check that the device is working properly. The installation is simple and the protection encompasses the entire branch circuit, so these devices are quite good but, unfortunately, they are costly, and for this reason their use is limited.

The most common GFCI takes the form of a duplex receptacle, available in 15- and 20-ampere ratings. These devices fit in a deep wall box and take a special wall plate that has a large rectangular opening. They have test and reset buttons and the newer ones have LEDs that indicate tripped status, which is a good feature.

These GFCIs also have feed-through capability. Black (hot) and white (neutral) wires from the source are connected to terminations marked LINE. Two other terminations are marked LOAD, allowing downstream non-GFCI receptacles to be daisy-chained as needed. These receptacles also have GFCI protection. The GFCI comes with labels marked GFCI PROTECTED that may be affixed to the downstream receptacles.

FIGURE 3-5
Ground-fault circuit
interrupter.

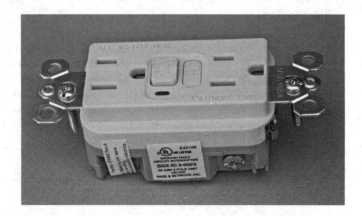

An interesting property of a GFCI is that it does not require an equipment grounding connection in order to function. In fact, a GFCI is a Code-sanctioned replacement for a receptacle that lacks means for an equipment grounding connection. Another type of GFCI is for cord-and-plug-connected portable electrical equipment that could present a hazard because it operates in a wet environment. An example is an electrically powered pressure washer. The GFCI is an integral part of the cord, close to the plug, and it also has test and reset buttons.

All of these GFCIs work the same—internal circuitry compares the incoming current on the black (hot) wire to the outgoing current on the white (neutral) wire. When the equipment and associated wiring are working properly, the difference between these two currents is zero. However if, due to faulty insulation or water infiltration, some electrons go to ground so that there is less current going back to the electrical supply than coming from it, a hazardous situation could occur and the GFCI trips out. Typically, a power tool such as a portable electric drill will develop loose bearings and the armature will rub on the housing. Alternatively, an internal wire, often where the cord enters the enclosure, will chafe and short against the metal. Either way, the housing will become energized. If it is grounded via a low-impedance path back to a good grounded service, the overcurrent device will trip, protecting all concerned.

However, if there is a weak link in the ground path (as when someone has sawed off the ground plug), the housing will remain energized. If someone takes hold of it while standing on a damp surface, that unfortunate individual will become the ground path and could be killed.

That is where the GFCI comes in. Sensing the missing electricity that is not returning to the power source via the neutral conductor, the GFCI instantly shuts down the circuit. These great lifesavers are Code mandated for many indoor and outdoor locations. "Residential" and "commercial" are not NEC terms. Instead, there is frequent reference to "dwellings" and "non-dwellings." The latter would include commercial and industrial locations, so the meaning is broad. GFCI requirements differ markedly depending on location. As an example, in the kitchen of a non-dwelling such as in a restaurant, all receptacles must be GFCI protected, whereas in a dwelling only the countertop receptacles and those within 6 ft of a sink must meet this requirement.

Nuisance Tripping

Because of their great sensitivity, there is a certain amount of nuisance tripping, and often the electrician is called in to solve the problem. It is one of the more common troubleshooting tasks. The procedure is not complex, but homeowners are perplexed and often need help.

There are three probable causes of GFCI tripping: faulty electrical equipment connected to a GFCI-protected circuit, faulty wiring associated with the circuit, or a faulty GFCI device. The fault has to be the device or downstream of it. Nothing that happens upstream can cause the GFCI to trip out.

The first step (after interviewing the homeowner for possible clues to the location of the fault) is to discover the extent of the affected circuit. To do this, hit the test button of the parent GFCI and find which downstream receptacles are dead. Look at the layout of the rooms with respect to the entrance panel or load center and think how you would run the branch circuit. This will help you to figure out which receptacles are fed by the GFCI. In large subdivisions where a few dollars saved here and there will affect the worth of the electrical contracting company over the years, it is common practice to bug an outdoor receptacle off a bathroom parent GFCI. This can be baffling for those who are not aware of this detail. In addition, it is possible that not all GFCI-protected conventional receptacles are labeled.

One at a time, unplug any pieces of electrical equipment to find if one of them is at fault. It could be as simple as a laptop charger or a gas stove with electronic igniter and clock. If it is not one of these, divide the string of paralleled receptacles in half by breaking the circuit at midpoint. You can disconnect the black wire only. See if the fault persists. Continue subdividing until the fault is found. It can be in the receptacle, in the wiring within the box, or in a concealed portion of the wiring between boxes. If it is in the concealed wiring, you have some work to do, either fishing new wire or running Wiremold. Such problems are the downside of GFCIs, but the lifesaving benefits outweigh such aberrations.

Just as GFCIs protect against electric shock, the more recent arc-fault technology shows great promise in the prevention of electrical fires. Electrical fires cause far more fatalities than electric shock does, so any innovations along these lines are certainly beneficial.

Electrical fires have various causes. Faulty design or initial installation can result in gross overfusing so that at a critical amount of loading the branch circuit wires become red hot within the wall. Neglecting to use corrosion inhibitor for aluminum connections at a service is another cause, and still another is overheating of a motor or other electrical equipment, perhaps caused by accumulation of dirt and debris that block air circulation. An arc-fault circuit interrupter prevents none of these.

Unlike a GFCI, an arc-fault device is sensitive to upstream and downstream faults. However, it only protects the downstream portion of the wiring. An arc fault is caused by a loose connection, partial break in the wiring, or other failure mode that is characterized by rapid and sharp variations in the current flow. It is a sputtering, buzzing sort of phenomenon, and the arc-fault device has internal circuitry that detects this aberration in the current that flows through it, immediately interrupting the power supply. Recent Code cycles have greatly expanded areas where its use is mandated in dwellings. If an arc-fault device is tripping out, avoid the temptation to swing over the supply wiring to a non-arc-fault device to restore service, as this leaves the latent fault in place, setting the stage for an electrical fire down the road. Troubleshooting and repair procedures are similar to those for GFCIs that are nuisance tripping. Sometimes a lot of work has to be done to clear the fault, but this goes with the territory.

A frequent homeowner complaint is high electricity usage as reflected in the utility bill, and here again an electrician is consulted. There are several ways the situation can be resolved. One answer, of course, is that an appliance is consuming an abnormal amount of power. As they age, motors begin to run slower and draw more current. Other types of electrical equipment can become big energy consumers as well, so once more we begin by entering an information-gathering mode. At times, the power-hungry appliance will be seen to function in a strictly steady-state fashion, drawing the same amount of current at all times. On the other hand, the consumption may be intermittent, fluctuating rapidly because of a poor connection, switch, or faulty heating element. Still another pattern of excess power consumption is that it is long-term variable. The current level will be time-dependent, with gradual changes keyed to an 8- or 24-hour period. This is caused by varying power usage of other electrical equipment that may be on the same premises or outside. Nevertheless, a fault in the unit under consideration is causing it to be vulnerable to this malfunction.

No matter the cause, measurements are in order. For steady-state excess current draw, the clamp-on ammeter is the perfect diagnostic instrument. If the appliance is alone on the branch circuit, you can find a single conductor to enclose in the jaws of the meter within the breaker box. If it is a hot water heater, the thermostat (temperature knob) and timer, if any, will have to be set properly. Record the ampere reading and verify that the thermostat shuts off the power supply after the temperature comes up. Of course, the whole problem could be a hot water leak or dripping hot water faucet. Compare the ammeter reading to the nameplate rated current and see if it is abnormally high. Electric hot water heaters usually have two elements of different wattage ratings. They are controlled separately by the thermostat and the setup can be changed by shifting terminations. Sometimes an element will burn out (become open) and this will cause the thermostat to feed power to the other element for an inordinately long time.

An Easy Job

An element resembles an incandescent light bulb but with a very heavy iron filament and no glass envelope. The element must always be immersed in water or it will instantly burn out. For this reason, usually the top element burns out. Either element can be checked in place with an ohmmeter. It is an easy job to change an element. With power locked out, drain the tank. Unhook the two wires from the element and unscrew it. If the gasket is damaged, it will have to be replaced. Hot water heater elements are inexpensive and generic elements fit various makes. Be sure to refill the tank before applying electrical power.

> If an incandescent light fixture will not work with a good bulb in place and power coming in, it is possible that the center spring terminal in the light socket has lost its tension. With power off, grab the "piton" with needle-nose pliers and give it a good pull. This often works on old fixtures.

Electrical equipment that is drawing intermittent or variable amounts of power can be monitored by means of a clamp-on ammeter that has a hold function. The meter can be left overnight or longer, to record peak power consumption. In this way, all appliances, branch circuits, and so on can be checked for inefficient operation.

Another cause of excessive electrical consumption is poor grounding. An inadequate ground can cause unstable voltage in the two legs of a single-phase service, and for a variety of reasons it should be corrected. The problem can be in the grounding electrode or the grounding electrode conductor.

Just about all residential services are grounded by means of two ground rods driven at least 6 ft apart. If they are closer, the two cylindrical-shaped charged regions in the ground interfere with one another, making for a weak grounding electrode system. The goal is to achieve a low-impedance connection to the earth. Besides ground rods, ground rings, ground plates, rebar in footings, and other grounding electrodes are effective, and specifications for them are given in the Code. Buried waterline, especially connected to a well casing, is an excellent grounding electrode, but unfortunately from the point of view under discussion the trend toward increased use of plastic pipe means water pipe grounding electrodes are no longer as reliable as they were at one time. You never know when a break in the pipe will be repaired with plastic, breaking ground continuity.

The NEC requires, when a ground rod is installed, that the ground resistance be measured and that it be not more than 25 ohms. However, this measurement cannot be made with an ordinary ohmmeter, as this would require a known perfect ground, and then there would be no point in the whole exercise. Ground resistance can be measured with expensive specialized equipment, but the Code, in an exception, provides that the measurement need not be taken if a second ground rod is installed, and this is how 90 percent of the residential services in new construction are built.

A certain amount of common sense is needed. Good grounding is dependent upon moisture content of the soil. Dry gravel in an arid region makes for a high ground resistance. The best plan is to place rods right in the drip line of the roof runoff. As mentioned, a 6-ft minimum distance between the two ground rods is required, but why not go farther for the price of a few feet of wire? The ground wire should be bare copper, buried below the drip line. The ground rods should be driven deep enough so that the tops, with ground rod clamps, are below grade. If bedrock is encountered, never saw off the ground rod. It may be driven at an angle or laid level as deep as possible.

If you suspect poor grounding is the reason for high meter readings, you can enhance the grounding system by adding extra electrodes, perhaps constructing a ground ring. In dealing with this problem, a utility engineer should be willing to visit the site for consultation.

These are some of the principal troubleshooting issues encountered in residential electrical work. Most of the troubleshooting in this setting involves an initial interview with the owner, the endeavor being to discover the exact nature of the problem and whether it was preexisting or arose in response to some specific event. It is surprising how often the user can supply a bit of insight that will lead you to a solution. Then, use the half-splitting technique to progressively narrow in on the defective component or device, and that should lay the foundation for a successful repair.

Commercial and Industrial Wiring

M ost of what we said about residential wiring also applies to commercial and industrial work, but the non-dwelling setting is typically much larger and contains an abundance of highly specialized electrical equipment and types of circuitry that are not seen in the home. Higher voltage and power levels are common, and where there are flammable gases or explosive dusts, many areas are classified as hazardous and require special wiring methods. A large manufacturing facility will have an in-house electrical department made up of professional specialists who possess considerable knowledge and expertise going way beyond what is required to wire a house. This is not to disparage residential electricians. Their work, though narrower in scope, requires impeccable attention to detail and the responsibility cannot be overstated. Moreover, the majority of electricians wear both hats in the course of their life work.

The greatest single difference between these two environments is in the dominant wiring method. For branch circuits, residential wiring is almost exclusively done in type NM, which stands for nonmetallic-sheathed cable. It is usually known by the trade name Romex.

Each type of cable and raceway is covered in National Electrical Code (NEC) Chap. 3, Wiring Methods and Materials, in Articles 320 through 398. Article 334 covers type NM and, as in all of these articles, there are subsections titled Uses Permitted and Uses Not Permitted.

Using Type NM

It turns out that type NM is permitted to be used in one- and two-family dwellings and their attached or detached garages and storage buildings. Type NM is permitted in other locations as well, depending upon the fire rating of the building. In actual practice, most commercial and industrial electrical work is done using type MC (metal clad) cable and EMT (electrical metallic tubing) and these are permitted in all locations except for the most restrictive of hazardous areas.

In a large facility where it is frequently necessary to test incandescent or self-ballasted fluorescent bulbs, you can make a quick tester mounted on the wall above your bench in the maintenance shop. Using a hacksaw, split the screw shell of a porcelain base fixture. Then, you can quickly stick in the bulb to be tested without screwing it in.

MC and EMT work well in commercial and industrial settings, and it is a simple matter to transition from one to the other—just terminate them through separate knockouts in a 4 × 4 box and wire nut the conductors. EMT is preferable for most long runs with few bends. In tight places where there are numerous bends, or through drilled holes in studs, it is easier to snake MC as needed. Then you can transition back to EMT.

EMT is technically not a conduit. It is tubing, although in using it most electricians talk about "putting the wire in conduit." EMT has a thinner wall than rigid metallic conduit (RMC), which resembles galvanized water pipe. EMT is not threaded in the field. It slides easily into fittings, where it is held in place with setscrews or, for wet locations, compression fittings.

For small jobs, EMT can be cut with a hacksaw. If there is a lot of work to be done or for larger sizes, a portable bandsaw is more efficient. All cut ends must be reamed to remove sharp burrs. Bending conduit starts very simply and quickly becomes complex. The basic premise are that bends must be gradual and uniform, so that the internal diameter is not materially reduced. To get a good bend, it is necessary to use a hand or hydraulic bender. These are easy to use, but it is essential to visualize the product and how it will interact with the building layout to get a good finish appearance. A fundamental principle in all conduit work is that the raceway must closely follow the wall or ceiling surface, and not take shortcuts at odd angles that would clutter the available space in the building (see Figs. 4-1, 4-2, and 4-3).

The Code, of course, regulates many aspects of raceway installation, and it should be consulted for information regarding minimum spacing of supports, maximum number of bends, and similar requirements. One basic rule is that the entire raceway system is to be assembled and terminated at both ends before pulling the wire.

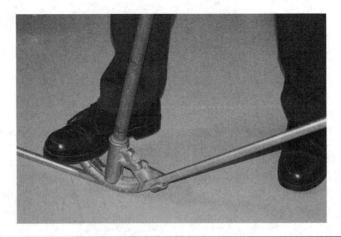

FIGURE 4-1   Bending electrical metallic tubing.

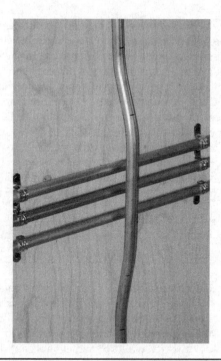

FIGURE 4-2   Four-point saddle bend.

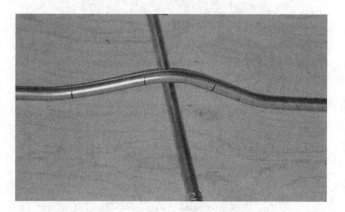

FIGURE 4-3   Three-point saddle bend.

Simple Conduit Bends

To begin with a simple exercise, suppose we are coming out of the top of a load center that is already mounted on the wall. We want to bring the EMT up to the ceiling, whereupon it will make a 90-degree bend and run along the ceiling to a fluorescent light fixture. You will notice that where the raceway exits the enclosure, it is spaced some distance from the wall. To follow the wall, the pipe needs to make two bends. In combination, they are called an "offset bend." This consists of two bends at the same angle, in the same plane, but in opposite directions, so that the pipe runs that are on either side of the sloping segment are parallel but at different elevations. The amount of offset depends upon the angles and the distance between them, and this is what you have to decide to make the offset. Center-to-center, the length of the raceway between the two bends is equal to the depth of the offset multiplied by the cosecant of the two angles. (Cosecants for various angles are given in trigonometric tables.) The most commonly used offset angle is 30 degrees, which is preferred by electricians because the cosecant, 2, is easy to remember and to calculate. For offsets less than 4 in., however, 30 degrees is too large an angle. Frequently used angles are 22.5 degrees, with a cosecant of 2.6, and 10 degrees, with a cosecant of 5.76.

> For a superior finish appearance, tighten screws in wall plates so that the slots are perfectly vertical.

For a standard box offset, a very useful tool is the box offset bender, which makes both bends at the precise angle and in the same plane with a simple push of the lever (see Fig. 4-4). After making the offset bend that is required so that the EMT conforms to the wall, we have to bend the pipe at a 90-degree angle so that it can travel along the ceiling. This configuration is called (by me) a "brick wall," by which is meant that the start of the

Figure 4-4   Box offset bender.

raceway is fixed (because the box was already mounted on the wall, and the adjacent building surface, a ceiling in this example, is fixed). To make this bend, measure the distance to the brick wall and deduct a certain number, depending on the size of the EMT. For ½-in. pipe, it is 5 in., and for ¾-in. pipe, it is 6 in. The number is stamped on the bender. Measure from the end of the pipe, deduct the correct number, and mark the EMT using a fine felt-tip pen. Line up the mark with a dedicated symbol on the bender, and make your 90-degree bend. The piece will conform to the wall-ceiling corner.

An easier method, if a little less elegant, is to start from the box with a short length of EMT, the one with the offset. Then make the 90-degree bend in a long piece, hold it in place to mark the length of the leg going down the wall to a coupling, and cut to fit. Another shortcut, where the exact height of the box is not critical, is to install the raceway first and then mount the box at a height conforming to the pipe installation.

What remains is to fit the piece needed to reach the light fixture using the box offset bender. Install an EMT connector with a locknut on the inside. Drive the locknuts at both ends very tight so that they scratch through the paint and dig into the metal, making a good grounding connection.

Now that the raceway assembly is complete, the time has arrived to pull the wires. For this indoor installation, 12 AWG THHN will probably be used. For a short, simple installation like this, a pull rope will not be necessary. Set up the three spools of wire— white, black, and green—on a rod suspended at both ends. A pair of stepladders works well, but for big jobs, you will want a portable wire caddy. Tape the wire ends together and push the wire through the raceway from the box to the fixture. When the wire comes to a dead stop, it has probably arrived at the light fixture. Trim the wires at both ends and terminate them. If there is a switch loop, follow the same piping and wiring procedure from the fixture or panel to the wall box.

This is a typical small EMT job. Some bending projects will be more complex, but industrial electricians do this work day after day, and it becomes familiar. EMT, as the word tubing implies, is the lightweight of metal raceways. However, it provides ample protection and shielding for common commercial and industrial applications. Intermediate metal conduit (IMC) and RMC are needed for hazardous locations and particularly rough environments. You are free to use them anywhere EMT is permitted, but the material cost is much higher and the labor—lifting, cutting, threading, and screwing together—for these heavier pipes is a very big factor, especially in larger sizes.

One benefit of any metal raceway, including EMT, is that conductors may be easily replaced or extra circuits added, provided the pipe is not overfilled, merely by pulling out old and installing new. Another advantage of metal raceway is that when grounded (which is always), it provides excellent electrostatic and radio frequency shielding for data and communications circuits. This shielding eliminates crosstalk, hum, and signal corruption. For ordinary Ethernet lines, although not mandatory, it is an excellent practice to install them in EMT between terminations.

> In an excavation or other damp place, use cordless power tools only. If this is not possible, make sure that the power supply is GFCI protected.

Regarding troubleshooting and repair, EMT offers some advantages but also presents unique challenges, particularly concerning tracing and identifying individual conductors.

If it is a Romex installation, it is often possible to follow individual cables in the basement of a residence where they are not concealed behind wall or ceiling finish. Where many wires of the same color are installed in EMT, identification can be difficult, especially if the original installer did not leave a good directory. There can be many 12 AWG circuits in a single large raceway, and there will be no way of telling which white (neutral) goes with which black (hot). This information is needed when doing certain types of upgrades such as adding arc-fault protection.

Identifying Wires

For short EMT runs, have a helper tug on the conductors at one end of the run and see if you can pin down the correct wire at the other end. Alternately, with a heavy load connected at the downstream end, remove the whites from the neutral bar one at a time and do ohmmeter tests.

Ungrounded conductors (hot wires) can be identified by using any colors except those reserved for the grounded neutral and equipment grounding conductor. For single-phase, 120-volt circuits, black is generally used. As an alternate hot wire for three-way switches or a two-speed motor, red is most common. When pulling multiple circuits in a single raceway, blue, yellow, pink, orange, brown, and so on are used for the hot wire, in addition to black and red.

Though not Code mandated, it is trade practice to denote three-phase legs as follows:

- Lower voltage, typically 208Y/120V: black, red, blue
- Higher voltage, typically 480Y/277V: brown, orange, yellow

In tracing circuits, there are some specialized tools that work well. In an industrial or commercial facility, there is often a substantial amount of telephone work to be done. New lines are needed from time to time, and maintenance is needed when something goes awry. The relevant tools are a toner, which injects a pulsed audio tone at one end of a line; a test set, which clips onto lines to listen for the pulses, talk to coworkers, or dial an outside line; and a wand, which when brought near a line to which the toner has been connected will pick up the signal. There is no reason why these tools cannot be used to trace de-energized power lines.

Type MC, as mentioned, is permitted in many of the same locations as EMT, and the two are often intermingled as part of the same job. The main difference is that for EMT, the metal tubing, properly terminated, qualifies as an equipment grounding conductor, while the metal armor of type MC does not. This is mostly a nonissue, however, because MC contains a green wire that is the equipment-grounding conductor. The only grounding advantage possessed by EMT is when redundant grounding is required, as in the patient bed area of a healthcare facility.

An essential tool for working with Type MC cable is the Roto-Split. It cuts through the metal armor without nicking the enclosed conductors. To make a termination, cut off approximately 8 in. of the metal sheath and slide it off. Cut back and remove the plastic filler strip. Put on the antishort bushing ("redhead") and attach the connector. Snap-on connectors are easy to use and do not require the bushing. Install the cable through a knockout in the box and wire nut the conductors. Use a grounding pigtail to ground the box.

Details concerning uses permitted, uses not permitted, spacing of securing hardware, minimum bending radius, and so on are found in NEC Article 330.

Eventually, many industrial electricians have occasion to work on programmable logic controllers (PLCs). The advantages of this technology in assembly-line manufacturing operations are numerous and there are millions of these installations worldwide. There are specific setup and programming protocols that are similar, although not exactly the same, among the various makes. You may find at first that there is a somewhat steep learning curve, but the background information is readily available from manufacturers, online forums, and print-format textbooks. After a little experience, most technicians agree that these systems are remarkably user-friendly and efficient.

PLC Anatomy

The brain of a PLC system is a modular, rack-based, central processing unit (CPU) that resembles your desktop computer in that it has memory and software. Systems may be as simple as a $200 "brick" that fits in the palm of your hand, to large, fan-cooled, floor-to-ceiling enclosures that orchestrate complex machinery. For initial installation and to make changes, the technician goes to the PLC control panel, plugs in a laptop computer, and programs the unit, using an on-screen ladder diagram that can be constructed and manipulated using the computer mouse and keyboard. The control panel is wired by means of low-voltage circuitry to actuators, motor controllers, sensors, and so on that are attached to the assembly lines and production equipment throughout the facility, sometimes in different buildings. The PLC gathers information and, based on its internal timer and programming, issues commands that result in an efficient workflow.

Most repairs involve the motors, sensing apparatus, and actuators rather than the actual PLC, although when first entering programming a certain amount of debugging may be required. PLCs surfaced in 1968 when General Motors embraced the technology in order to improve upon antiquated systems of switches and relays that had to be rebuilt once a year when new car and truck models appeared. Instead of hardware elements, PLCs have programmed instructions that reside in microchips on printed circuit boards within the CPU.

What quickly evolved was a rack-based system that accepts input and output modules. They may be digital or analog and plug into slots in order to accommodate whatever sensors and actuators are required by the installation.

The PLC consists of a very rugged housing that can withstand vibration, dust, moisture, and other environmental stresses. Within the housing is the solid-state circuitry that initiates actions as needed to perform the tasks. Disc-based software permits the technician to insert virtual components onto the on-screen ladder diagram to program the PLC. The two ladder rails are positive pole (left rail) and ground bus (right rail). Programming begins at the top of the ladder and each rung is made up of an input and an output, one complete executable command. Each of these has a numerical address, such as 0500 or 1000. An example would be a rung with a single-pole switch on the left and a lighting load on the right.

An operating PLC is continuously scanning. The PLC looks at the input to ascertain whether it is ON. Then, the PLC executes the programming, beginning at the top of the ladder. Finally, the PLC updates the output status. There are two modes—programming mode and operating mode.

To make it all work, input and output modules are inserted into the slots. Examples of inputs are limit switches, pressure transducers, voltage sensors, flow meters, and thermocouples. The output modules are wired to low-voltage control circuits to motors and other actuators. An internal oscillator-based clock permits the PLC to scan the inputs. Latching or momentary outputs permit the machinery to function in a harmonious and productive fashion.

Maintenance electricians who work in a facility having PLCs have an easy way to get into the field. Manufacturers, such as Allen Bradley and Siemens, have free online material including machine-specific documentation, so there is a clear shot at gaining on-the-job expertise in this interesting field. Two good introductory websites are www.plcs.net and www.plcdev.com.

To effectively troubleshoot and repair PLC systems, the technician will need good electrical, computer, and mechanical skills. The undertaking is somewhat more advanced than most electrical work encountered on a routine basis, but then that is what makes it interesting.

The maintenance electrician is usually the first responder when PLC-controlled equipment fails to perform as expected. Often it is an incredibly simple fix, but there is always the potential for problems to spread and multiply, particularly if a deranged system has caused mechanical jamming or worse on the assembly line.

As always at the outset, conduct a thorough interview with the operators. Is the equipment completely dead, is it tripping out instantly if power is applied, or does it trip out after a period of time?

The PLC output voltages are low control voltages that activate, let us say, a motor controller, so it is important to see if there is line voltage at the input and, with contacts pulled in, at the output. If so, is there voltage at the motor? Do the usual motor and driven load checks, exercising extreme caution that you do not start the motor where it could endanger workers who are within range of driven machinery.

If the problem is not in that area, you can work back toward the PLC, or start at the upstream end, depending in your judgment which of these approaches is more likely to take you to the fault. From the alphanumeric display and the power light, it will be evident if the PLC is receiving power. If not, check the breaker and branch circuit wiring. If the supply power is 240 volts, make sure that both legs are intact.

If the schematic is available or from inspection, note whether the 24-volt dc power supply is inside of the unit or supplied externally. There may be primary fuses that should show no voltage across them if they are not blown.

There should be internal diagnostics, depending on the make and model of the PLC. If an error code is given on the alphanumeric display, look it up in the PLC documentation or consult an Internet search engine, as these resources usually provide the answer.

Look at the input LEDs and manually activate the switches. For this operation, if there is a conceivable hazard, be sure to have any motors or actuators powered down and locked out.

If the inputs are all good, put your meter on the outputs, and this will reveal the status of the PLC. Check if the proper voltages are coming out of the PLC so that all solenoids will turn on as needed. If there are relays that make the outputs 120 volts, one of these may be stuck or have an open coil. See if the relay pulls in as required.

These are the basic diagnostics for a PLC, but details will differ, depending on the make. There is usually good technical assistance available from the manufacturer, and sometimes the distributor will have someone with a wealth of information to share.

In industrial wiring, the overriding issues that confront electricians every day are complexity and integration, and these considerations make it markedly different from residential electrical work. An industrial or large commercial facility will have a great many systems and subsystems, and they are frequently connected in one way or another, so to competently maintain, diagnose, and repair the equipment, workers need to understand how these elements work together. They need an extensive knowledge of each piece as well as an understanding of how it relates to the whole.

A manufacturing facility, to take one example, typically has elevators, sprinklers, a fire-alarm system, a telephone system with in-house private branch exchange (PBX), and other related systems.

If a sprinkler head ruptures due to heat from below, the water flow is sensed and this activates the fire alarm system. The fire alarm control console initiates a call to the fire department or monitoring agency on one of two dedicated telephone lines, and it causes elevator cars to proceed immediately to the ground floor or, if fire is detected there, to some other predetermined floor. Additionally, LP or natural gas, exhaust ducts, and other systems can be shut off or redirected as mandated by local codes and building requirements.

In the remainder of this book, we shall look at some of these systems as discrete entities and also as parts of the larger structure, particularly as troubleshooting and repair procedures relate to them.

Electrical Power Distribution Systems Including Transformers

Thomas Edison and his associates were incorrect in thinking they could build a large-scale electrical distribution system that would work on direct current. The only way their dc system could function would be if it consisted of a huge number of small neighborhood-based power plants with large diameter copper conductors providing power to the individual buildings, not to mention streetlights and other loads. Moreover, there could be no such thing as three-phase power.

Electrical power is the product of amperes and volts. This means that for a given resistance in a length of power line and the load, if the applied voltage is increased, the amperage will decrease. Since the ampacity of wire is measured by the amount of current it can carry without overheating, it is clear that electrical power can be transmitted more efficiently at higher voltage levels. Therefore, the strategy will be to use higher voltages for long transmission lines, dropping the voltage to a more useable level at the destination.

The only ways to change dc voltage levels are electronically and by means of motor-generator sets, neither of which is feasible for numerous high-power applications. AC voltage, on the other hand, can be changed efficiently and relatively inexpensively by means of transformers. The moving magnetic flux that is necessary for inductive coupling is created not mechanically but because of the changing voltage polarity (and consequently current direction) in the primary winding of a transformer. Large amounts of power can be stepped up or stepped down merely by varying the ratio of the number of primary to secondary winding turns.

Consequently, Tesla and Westinghouse prevailed, and today enormous ac electrical grids power the world. With a rated capacity of 830,000 megawatts, the North American Grid is the largest machine ever built. It provides electrical power for most residences and businesses in the United States, Canada, and parts of Mexico. A major design goal is reliability, and this is achieved by pooling the power from numerous generating plants. The electricity combines to make a single output for each of the three separate sectors of the North American Grid, in which the ac is synchronized and phase locked at 60 Hz. The three sectors—the Eastern, the Western, and the Electric Reliability Council of Texas—are linked by huge dc interconnects, so that when a single power source goes down, others make up the deficit. Considering the size and complexity of the Grid, widespread outages are few and, for the most, part brief.

> The first step in servicing many types of nonfunctioning tools and appliances can be to fasten an ohmmeter to the plug. Stick the probes through the holes in the prongs and hold them in place with a wrap of electrical tape. Put the ohmmeter in honk mode and work the power switch or wiggle the cord. For a motorized tool such as a drill, you need to have continuity through the cord, switch, and associated wiring, and through the brushes all the way to the armature.

To understand how these connections work, we must look at the way in which two or more ac generators can be synchronized. To begin, ac power sources must be the same voltage and frequency. Furthermore, the phase sequences and angles must be identical. Then, with the two sources electrically isolated (breakers in the OFF position) the speed and hence frequency of one of the sources is adjusted until the waveforms of the two coincide, at which precise moment they are placed together online so that they are connected in parallel. Once they are so joined, the smaller power source tends to remain in synch with the larger one, and useable power is available from the combination.

Problem Solving

In this way, all ac power sources within each of the three sectors of the North American grid are phase locked, although the three sectors are not synchronized with each other because they are connected only by dc lines.

Some power sources, such as utility-scale photovoltaic arrays, produce dc output, which has to be changed by means of a synchronous inverter to ac of the proper voltage and frequency so that it can be synched into the grid.

Each subcircuit within the grid is composed of three parts: generation, transmission, and distribution. Providing the links between each of the stages and between elements within them are the ubiquitous substations that we see as we drive along roads anywhere in the world (see Fig. 5-1).

In addition, many substations are invisible to us because they are within buildings or underground. When substations are utility-owned and operated, they are not under National Electrical Code (NEC) jurisdiction; however, if they are privately owned within a large industrial complex that includes power transmission and distribution, then they are subject to NEC mandates. In either case, the design goals are essentially the same: transformation, short-circuit and overcurrent protection, switching, monitoring frequency and other electrical parameters, metering, lightning protection, capacitors for power factor correction, lighting, alarm, communication, etc.

Transformation is usually of three-phase power, often accomplished by an array of three transformers (see Fig. 5-2), or in the case of an open Delta, two transformers (see Fig. 5-3). I like to stand outside a substation enclosure and see if I can visually trace the wiring. This may seem like a strange affinity, but perhaps no more so than playing bingo or attending a rock concert.

FIGURE 5-1   Substation with grounded fence.

FIGURE 5-2   Three-phase transformer array.

When putting in a new service, put a very small dab of corrosion inhibitor on each terminal of the meter socket. This makes it easier to insert and remove the meter.

Short-circuit and overcurrent protection are overriding concerns. The principle is the same as in a small-scale service, but the breakers are very expensive, costing thousands of dollars. A difficult design objective is extinguishing the arc. If voltage and current are high, as the contacts open electrical energy will continue to flow between them because an ionized path through the air is established, becoming hotter and more intense even as the

gap increases. This is essentially what happens in a lightning event, and various strategies are employed to extinguish the arc before it burns up the equipment. Some possibilities are compressed air pressure, oil or sand immersion, magnetic deflection, and mechanical intervention.

Concerning switching, it must be said that this capability is essential within a substation. It is necessary to isolate and de-energize discrete portions of the load so that repairs or new construction can be done without causing a widespread outage.

Metering is also essential within any substation. It is an important troubleshooting tool in attempting to pin down load irregularities. As customer loads become increasingly nonlinear, harmonics problems tend to surface, causing unwanted heating of the neutral. Effective metering can help resolve such issues.

Lightning protection has progressed rapidly in recent years, and it has grown in importance as power lines become longer, with more customer outlets. There is a lot we still do not know about lightning, including the exact mechanism for the formation of charge clusters within clouds, and what triggers the release of such awesome quantities of electrical energy. Nevertheless, researchers have come to understand that damage to electrical equipment can be prevented by minimizing bends in transmission lines and, above all, creating low-impedance grounding at key locations, both utility- and customer-owned.

Capacitors may be placed online as needed for power factor correction. They function to compensate for large inductive loads that cause current to become out of phase with voltage.

All of the substation properties, plus internal wiring, alarm, and communication must be documented and mapped out to facilitate new construction and upgrades.

SCADA and Remote Monitoring

Because substations are usually not maintained by fulltime on-site workers, there has to be a mechanism for remote monitoring. This is accomplished by Supervisory Control and Data Acquisition (SCADA). It is used in many applications including gas and oil refineries and pipelines, water treatment plants, wind farms, airports, and so on.

The substation can transfer data to the monitoring facility when solicited or on its own initiative, so this protocol is well suited for sensing as well as control. Numerous SCADA software programs are offered, and they can be simply installed in computers at the monitoring facility. These programs include provisions for PLCs, human interface capability with user-friendly graphics, and other bells and whistles. Often it is possible, when trouble arises, to take remedial action electronically without dispatching workers to the site. Alternatively, if it is necessary to send a crew to the substation, they will know what tools and materials to take with them.

For substations as elsewhere in electrical systems, grounding serves more than one purpose. Principal benefits of good low-impedance grounding are as follows:

- It mitigates the harmful effects of lightning, which besides damaging substation equipment can create a voltage surge that will travel throughout the distribution network to impact customers' electrical equipment and threaten buildings and the people within them with the twin demons of electrical fire and shock.

- It makes for a reliable return path to the electrical supply to facilitate the operation of overcurrent devices.

- It stabilizes voltages so they will not float over or under the design standard, either of which will damage motors as well as sensitive electronic equipment.

Where there is an electrical distribution system of any significant size, there will be three-phase power. For many nonelectricians, the idea is foreign and incomprehensible, mainly because it is not part of the ordinary residential setup. Moreover, most backyard mechanics, by and large a knowledgeable bunch, are surprised to learn that the familiar automotive alternator is in truth a three-phase generator, with an internal diode network that outputs a dc voltage appropriate for charging the car battery as well as powering the onboard computer and other dc loads.

> Telephone servicing is much easier with three simple tools: a test set, tone generator, and wand. Some technicians use an ohmmeter instead, but the foregoing reveal sound quality issues as well.

For electricians and utility workers, the good news is that three-phase connections are simple to understand and, on a comparative kilowatt basis, actually easier to install because the wire sizes are smaller than for single phase. Troubleshooting and repair are straightforward—just look for the dropped phase.

Wye and Delta

Three-phase power begins at the generator, and its unique characteristics are due to the rotary nature of that power source. The three generator windings are spaced equally around the rotating armature or field, creating three outputs of the same frequency and voltage. There are six wires emanating from the three coils, but within the generator housing or immediately outside, they are connected to make a three-wire or four-wire output. They are connected neither in series nor in parallel, but in either of two unique configurations known as Wye or Delta. Each of these has particular advantages for the supplier and end user.

In the Wye configuration, one end of each winding is connected to a common terminal, which is usually grounded and becomes the neutral. The other end of each winding becomes one of the three hot legs. Four wires make up this power supply, and single-phase power (using one of the legs and the neutral) and three-phase power are available in the same entrance panel.

The other somewhat less common but nevertheless very useful three-phase configuration is the Delta, so called because its schematic representation resembles the Greek letter Delta, an equilateral triangle with its apex at the top.

The center-grounded Delta has one winding center-tapped and grounded. The corner opposite the center-grounded winding is called the high leg because it has a higher voltage to ground than the other two legs. The high leg is always color-coded orange.

In another version of the Delta configuration—the open Delta—one winding is omitted, but there are still three hot legs. Because of this "virtual winding," the three voltages are present but the overall capacity is reduced.

If three single-phase transformers are used in a Delta bank to provide stepped-down voltage for customer use, and if one of the transformers fails, the configuration becomes an open Delta and continues to function with no outage on a temporary emergency basis, although at reduced capacity of 57.7 percent of full power until a replacement can be installed.

Three-phase power is 150 percent more efficient than single-phase power. Conductors are 75 percent the size of single-phase conductors. Motors, raceways, and other components are also smaller, so there are savings in material and labor. A problem that is encountered especially in rural areas, however, is that three-phase power is not always available from the utility, and costs for a line extension may be prohibitive for all but the largest facilities.

Eventually, there may be a move back to dc power, and Edison will have been vindicated, although in a way he never conceived. This is because of the enormous potential for photovoltaic power, produced locally, that could revolutionize the way electricity is brought to the home and business. At present, there are three models, with subdivisions and variations.

- Stand-alone systems are primarily seen in remote, off-grid locations. They are used for rural residences, where it is not feasible to bring in a utility line, and for isolated research facilities and data collection posts with environmental instrumentation and communication capability. Stand-alone PV systems are ideally suited for spacecraft where solar radiation is intense and continuous and the ambient temperature is low.

- Cogeneration systems are used where there is the potential for utility connection. The cost of a synchronous inverter is generally less than the cost of a battery bank.

- Utility-scale solar PV generation is seeing widespread use and as the price of solar cells continues to drop, it is realistic to expect that these installations will proliferate.

Solar technology falls into two main categories—crystalline silicon and thin film. With advances in manufacturing methods, the price per installed kilowatt is falling rapidly, making both of them increasingly attractive.

If sunlight strikes a material that is neither completely reflective nor completely transparent, there will be increased particle activity within it, in proportion to the amount of energy absorbed, that is, not reflected back toward the sun and not transmitted. This will be seen as a temperature rise in the material, but there will be no useable electrical output unless the material is a semiconducting photosensitive diode with leads attached.

As we have seen, a diode consists of lightly doped silicon that has a p-n junction. If leads are connected to both sides of the cell, there will be a measurable voltage and if a load is connected, there will be current flow as long as photons strike the surface of the cell and are absorbed.

Electrodes with leads must be attached to extract electricity, and this is not a problem on the bottom surface, but at the top where the solar radiation has to enter the cell, a simple metal plate is not possible because it would block the incoming light. A metallic grid must be constructed, its parts of sufficient mass so that the current will not be limited too much, but also not shading out an excessive amount of the cell's area. A strategy that has been successful is to make the conductor with a rectangular cross-section and place it so that the greater dimension is perpendicular to the surface of the cell so that sufficient light can get through and yet the conductor will not be overly resistive due to a small size.

The cells are wired in series-parallel configurations, creating panels and finally a complete array with the desired voltage and current capability.

Most electrochemical activity becomes more robust at higher temperatures, but PV generation is the opposite. It is stronger by a considerable factor when the ambient temperature is lower. For this reason, solar technology is suitable for cold northern areas where there may be less sunlight but colder daytime temperatures. Many people assume that intense direct sunlight is necessary for PV activity, but in fact, there is considerable available energy on a cold, cloudy day.

Code Navigation

For most of us, it is not possible to memorize the provisions of the NEC. Instead, learn to access answers to questions on an open-book basis. It is suggested that you spend a lot of time reading and understanding the table of contents and the index so that you know which of these to consult in various cases, and what the keywords may be. You should also know the contents of particularly important articles such as 240, 250, and 430.

The other main category of PV generation—thin-film technology—operates on the same fundamental principles, but the details differ. The silicon is laid on a metal, plastic, or glass substrate by the action of vapor deposition, followed by placement of a transparent conductive oxide coating that constitutes the topside electrode. The thin-film cell has good characteristics for the collection of light and export of electrons.

Thin-film technology, also known as amorphous silicon, is somewhat less efficient than crystalline silicon, but it is cheaper to manufacture. Kilowatt for dollar, it may be the better deal, depending on the vendor.

At present, thin-film solar PV comprises only about 20 percent of the market, but in a highly volatile environment, it could surge ahead. What is interesting, from the point of view of the builder, is that thin-film technology is well suited for building integrated photovoltaics (BIPV). This is because, as the name suggests, thin-film solar PV products, in some versions, are flexible and may be attached to either curved or flat surfaces that form the facades of contemporary commercial buildings. In BIPV, the thin-film array is not mounted on the roof with standoff hardware, but it actually becomes the roof, siding, or windows.

In the not distant future, we may see as a norm use of BIPV to supply on-site dc power. It is perfectly feasible to manufacture appliances and lighting that operate on dc, so the entire concept of a power distribution network could come to be eclipsed by autonomous stand-alone power generation.

For troubleshooting and repair of any solar PV system, thin film or crystalline, special considerations and techniques come into play, particularly on a commercial scale where there are higher voltage levels and greater amounts of available fault current.

Solar Difficulties

Of course, an individual solar cell is a sealed unit, wherein the conversion of radiant energy to electricity takes place on a submicroscopic level, so internal repair is not possible. However, in many situations, one or more cells could become open or shorted, making the array in varying degrees dysfunctional. Other failure modes could involve storage batteries, inverter, utility connection, system wiring, or problems in downstream premises wiring or loads. Troubleshooting techniques—user interview, analysis of symptoms, use of documentation and technical assistance, and half-splitting circuit sleuthing—all play a role in locating and fixing the problem.

However, where solar PV systems differ is that the system must be considered always live on both sides of the disconnecting means, even if it is locked out. This is so even while the array is covered by an opaque material (a cumbersome method at best) because of the possibilities of unexpected voltage due to system capacitance and backfeed from battery storage or utility power. Beware also of the backup generator that, sensing an outage, could come online unexpectedly. On a large commercial site, such a generator could be too distant to hear when it roars to life.

To begin, check input and output voltages at the inverter. If the system appears dead only at the downstream end, it is likely an inverter malfunction, and if measurements indicate this is so, you will need to look inside the enclosure. Using manufacturer schematics and documentation, trace the power flow, looking for any damage that is apparent visually, such as blown fuses or burnt or loose wires. Do a thorough cleaning and check for any airflow blockage that would inhibit cooling.

If there is no dc supply to the inverter, the next place to look is on the roof or other array location. Check wire nut or crimp connections within junction boxes, as they are often the problem due to moisture infiltration. If a single part of the array is found to be defective, look for blown fuses or circuit breakers that may need to be reset. A roof-mounted array is a natural target for lightning damage, and this is not always visually apparent.

Continue to work upward into the array and it should be possible to determine the cells or connections that are faulty. A frequent cause of voltage loss within modules is dust or pollen buildup on the upper surface. Also, be aware that over the years trees will grow, causing unexpected shading. This will first become apparent in spring or early summer when the trees are leafing out or there is new growth. New construction can also cause increased shading.

In the course of upgrading a service or reworking existing wiring, it is often found that the panel directory is in poor shape due to inaccuracies, bad calligraphy, or tattered condition. You can print a new one on your computer using heavy paper cut to fit. Use the Excel program to make rows and columns. If you do not use Excel, you can use Microsoft Word, but it is more difficult to line up the spaces for double-pole breakers. After doing one of these jobs, save the directory in your computer. Then, the next time you can just enter new data. Be sure to include your logo with phone number.

An entirely different problem in a PV system is with the load. Breakers, fuses, or switches could be at fault. If an overcurrent device cuts out, remember there may be an underlying cause. Motors often have internal fusing, even imbedded in a winding, or an open coil or wiring fault could be the culprit. An easy test is to replace the motor with another load and note the result.

Poor grounding and faulty neutrals can create difficulties that are sometimes difficult to locate. Keep taking voltage and current measurements at various accessible points and soon you will prevail.

As part of the customer interview, determine if additional loading has been brought online, as this may cause poor system performance and may require an inverter, battery bank, wiring and overcurrent protection, or solar array upgrade.

Many of the above troubleshooting guidelines are equally applicable to wind power and fuel cell systems.

Telephone Systems

For centuries, a nonelectrical acoustic device, like children make from two tin cans and a string, was known and worked surprisingly well for medium-distance voice communication. It bears the suggestive name "lovers' telephone" and was the ancestor of the acoustic devices we are so fond of today. From that pleasant diversion and the astonishing success of coast-to-coast and later transcontinental telegraphy in the nineteenth century, came the impetus for a worldwide high-quality telephone network, with perhaps five billion connected outlets including cell phones. A few decades ago, you had to shout to make yourself heard in a coast-to-coast call. Now, two or more people can speak via conference call to China and the sound quality is the same as being next door.

Alongside all this technical progress, the in-premises part of the infrastructure has become accessible through a process known as deregulation. The original business model had the utility owning all the equipment including the in-house wiring harness and the phone itself. Utility workers did the original installation plus any subsequent repair work, regardless of whether it was a modest cottage with one jack or a large office building with many individual lines and extensions and an in-house switchboard.

> To bend and shape large (1/0 and up) insulated conductors in order to make meter socket and panel terminations, use the hole, with its well-rounded edges, at the end of a large adjustable wrench handle.

Generally, the network installation and wiring were satisfactory, but the work from the point of view of appearance left much to be desired. If phone extensions were to be installed in many rooms in a house, the cabling was spider-webbed from a box on the street side of the house by means of tacking the lines to the siding outdoors and drilling in to the location of each phone. For splices, wires were twisted together, taped, and left to withstand the elements. The securing hardware was spaced far enough apart to ensure that the wires would flap in the wind and sag between supports. When repairs were made, abandoned lines were left in place for the life of the building.

Now, all premises telephone wiring has become privatized. In a commercial building, in-house electricians or maintenance personnel do the cabling, or the work is hired out to a cabling contractor or outside electricians who include such tasks in their job description. Overall, a much higher degree of professionalism prevails. If an extension is to be added in an office, the cable will be fished behind paneling or whatever it takes to conceal it. For factory work, a new run of EMT will be added, rather than hanging the cable from an existing unrelated raceway, which is a National Electrical Code (NEC) violation.

In troubleshooting and repairing premises telephone systems, it is important to maintain a high standard of workmanship, and to strive to create a product that, in addition to being reliable for years to come and convenient to use, will have a good, simple appearance without any unnecessary wiring or hardware that would be perceived as clutter.

If your troubleshooting and repair will involve any alteration or addition to the existing installation, it is vital to refer to the current NEC edition and to adhere to any applicable mandates. Previously, we outlined NEC requirements for data and communication wiring, specifically removal of abandoned cable and the cable hierarchy, which permits substitution of plenum cable in riser applications, but does not permit substitution of riser cable in plenum applications, to give a typical example.

PBX and Multiplexing

In a medium-sized commercial setting such as an office building with 500 lines, there may be a private branch exchange (PBX) on the premises, and it is interesting to see how it works. If the commercial site is a complex of buildings, it is likely that the electrical equipment that makes up the PBX will be housed in a small concrete building with no windows (to deter vandals). This is known as a central office. Into this building via a large diameter cable with typically 75 pairs of telephone wires come voice and data communication with the outside world. How is it possible that this trunk line is able to serve a system with as many as 500 lines? The answer is through the miracle of multiplexing.

This is an umbrella term for various methods for combining several analog or digital signals into one signal that can be transmitted in real time over a single line. Without going into the details because this knowledge is not relevant to troubleshooting phone lines within the PBX or in premises telephony, here is a list of multiplex methods:

- Space division
- Frequency division
- Time division
- Polarization division
- Orbital angular momentum
- Code division

The PBX contains equipment to demultiplex these incoming calls and prepare them for distribution throughout the branch lines or the reverse. The heart of the PBX, and the entire phone system, is the switch, which takes the place of the old human-operated switchboard that predated the dial telephone system. A central office from years ago was a noisy affair, with many operators simultaneously speaking to customers or, later, mechanical relays clacking in response to dialed calls from all parts of the exchange or in response to incoming calls from anywhere in the world.

A modern central office is eerily silent except when an alarm sounds. The switch is entirely silent. It is electronic, and within its protected environment are one or more rows of floor-to-ceiling enclosures. It is quite reliable. For each branch line, there is a dedicated slot within an enclosure containing a printed circuit card. These occasionally become defective, and can easily be swapped out, using the normal procedures to avoid static electricity-caused damage in handling. From there, the individual lines emerge to a cross-connect, where local lines are connected. They merge into one or more large, multipair cables that go to a centrally located cross-connect panel typically on separate floors in a building or in separate buildings.

The switch is computer-controlled from within the central office, in the premises maintenance office, or off-site via an Internet connection. Computer access to the switch is password protected, and there is the capability to control from anywhere in the world the many parameters of the system as a whole in addition to the properties of each user line. Telephone numbers, methods for accessing an outside line, call waiting, and caller ID are a few of the attributes that are accessed and controlled in this way. The details and full range of capabilities vary among different manufacturers, but what they have in common is multivolume print and online documentation with extensive error-code and troubleshooting information. Tech help is also available.

Large electrolytic capacitors have a limited shelf life. In the case of a good one, an ohmmeter set on a high range will "count" first one way and then the other as the leads are reversed and the internal meter battery alternately charges and discharges the capacitor. Sometimes a bad electrolytic capacitor can be brought back to life by applying a steady dc voltage for a long time.

A large commercial or industrial complex with an on-site PBX will need at least one individual, maintenance electrician, or telecom specialist who has obtained the knowledge and expertise necessary to operate the computer-controlled switch in order to troubleshoot and repair lines as needed.

Reliability is a high-priority design goal for the switch and associated equipment. To this end, the central office building must be carefully designed and built. Extensive grounding with a rugged, low impedance, conduit enclosed grounding electrode conductor is essential. Ufer technology, which means concrete-encased metal in contact with the ground, makes for an excellent grounding electrode, far better than the usual ground rods used in many types of construction. However, this method will not be effective if there is plastic or foam under the concrete, or if the rebar is covered by paint, epoxy, or other insulating coating. Other possibilities are a copper ground ring, plate, or mat under the building, or any of these in combination, possibly with additional ground rods. Regardless of the grounding method, a low-impedance connection to earth is desirable for a variety of reasons, particularly lightning protection, voltage stabilization, and protection from ac hum in the phone system.

An ordinary electrical service will be brought to the central office, but it is important that the telephone system functions without interruption during a power outage, regardless of duration. This is possible because of a large battery bank inside the central office. It is always online and receiving a maintenance charge from equipment that is connected to the utility supply. A highly visible digital voltage readout is connected to each cell. Proactive cell replacement, cable, and connections inspection and servicing should be part of the maintenance procedure. Additionally, there must be a backup generator, located immediately outside of the building and situated so that exhaust fumes will not get into the building's air intake. The generator is usually LPG or diesel. It starts automatically in the event of an outage, and takes over battery charging and building services. The automatic transfer switch ensures that there will not be backfed energy into the utility lines, which would endanger line workers. Periodically, the backup generator starts automatically and runs for a few minutes. Maintenance should include changing oil, inspecting belts, and checking the condition of the starting battery, including battery cable connections, as well as coolant level and condition. For these procedures, refer to the owner's manual.

> When soldering electrical work, use resin, not acid flux, which is good for water pipes and radiators, but will leave a corrosive residue on printed circuits boards and the like that will eventually compromise the conductivity of an otherwise properly soldered joint.

The central office requires other features in order to provide reliable phone service throughout the exchange. There has to be very reliable air-conditioning. This goes way beyond worker comfort. It is necessary to prevent temperature rise due to the large amount of heat that is generated by the switch electronics, lighting, battery charging, etc. It has to be sized out in conjunction with the highest anticipated outdoor ambient temperature. In addition, there must be reliable heat adequate for a prolonged winter cold spell. The temperature inside can be measured and automatically reported to the facility maintenance department, using inexpensive hardware. Keep in mind that excessive heat, due to failure of the air-conditioning, could fry over $1 million worth of electronics and make for a prolonged and widespread communications outage. Lighting must be adequate for exacting repair tasks that could become necessary. Needless to say, the roofing and structural integrity of the building must be sufficient to ensure that the large capital investment is not at risk.

We have outlined a relatively simple telephone system with PBX. If you have assimilated the switch documentation, including learning command-line entries for setting switch and individual line parameters, you know how to make keyboard queries, and you know the cabling layout, then you will be equipped to troubleshoot and maintain the telephone system and to perform alterations and additions to it as needed for day-to-day operations.

Phone Diagnosis and Repair

Troubleshooting a phone system is a little different from other types of electrical work. In some ways it is easier, although without the right tools and orderly work methods, some tasks can become overwhelming. There is much specialized equipment for telephone diagnosis and repair, but only a few simple tools are needed to get started. Besides ordinary electricians' tools such as a good multimeter, fish tape, hole saws, long electricians' drill bits, a complete set of hand tools and cordless drill/screwdriver, the following specialized telephone tools will be required:

- A telephone test set (butt set) is essential for almost every repair job. This is a compact phone with built-in line-powered electronics and keypad dialing. Leads with alligator clips are included. Simple models are available for under $100, but prices escalate if you want advanced features.

- A toner is an audio-frequency generator powered by a 9-volt dry cell. It has two leads with alligator clips or a short line cord with modular plug. You can inject the tone into a telephone line at any available point and listen for it elsewhere with a good connected phone or butt set. It is the preferred method for tracing and identifying lines with no helper required. You can attach it at the jack and work back toward the central office, indoors or outside. On the other hand, you can attach it at the central office and work downstream.

- A wand is a simple, noncontact (proximity) probe that allows you to find the pair carrying the tone merely by bringing the tool within a couple of inches of it. You can pass over large bundles of pairs quickly and know by the intensity of the tone when the extension has been found, so this device is a great time-saver.

- Some jacks and blocks have specialized terminations. The wires are not stripped, but "punched down" with a spring-loaded tool, and the terminations at the device pierce the insulation. The result is a good electrical connection that is sufficiently strong for the purpose. You can make these connections with needle-nose pliers, but the results are not consistent. You really need a punch-down tool. The only problem is that there is more than one style of termination, and each takes a different punch-down tool or different blades. The different connector types are 66, 110, BIX, and Krone.

- A wire-wrap/unwrap tool may be needed to work on the cross-connect frame within a central office. This is used for many types of electronic work. A detailed instruction manual comes with the tool, and it makes a very neat, durable connection with excellent electrical properties.

These are the principal tools needed for telephone branch line work, so it is inexpensive to get into the trade on this level.

Dead Phone Syndrome

Frequently, the electrician or telecom technician is called to check out a phone that is dead or noisy. Obviously, you need to find out whether the phone or the line is at fault. Pick up the receiver and see if there is a dial tone. If not, wiggle the modular connections at both ends of the receiver cord and line cord. If there is a response, most likely the fault is in the cord, which should be discarded and replaced. If there is no response, you will need to go deeper, but it is premature to rule out these cords. The best quick check is to plug a known good phone into the modular jack. If there is now a dial tone, and the good phone is appropriate for the location, you can leave it at that and the repair is complete. Take the bad phone back to your shop for further tests. Alternatively, you can try swapping out receiver and line cords, or just the receiver.

If there is still no dial tone, pull the cover off the jack and check terminations and the incoming line with your butt set. If it is a two-line setup, try the other line. Often, in seeking to make a dial-up Internet connection, someone will plug the phone into the wrong line connector, which may have been discontinued with the advent of wireless, and this is the reason for the "dead phone."

A shorted line or equipment (including computer, fax, or credit card terminal) may be plugged into a jack in the same room or nearby, or one of the lines or jacks may become shorted. A short circuit anywhere on a daisy-chained line will make for a dead line and phone. If this sort of problem is suspected, the best course of action is to disconnect all the phones and other equipment that are on the line and see if that brings back the dial tone. If it does, you can plug them in one at a time to find which one is at fault. If this does not bring back the dial tone, go to the first (farthest upstream) jack on the line, and check for the dial tone. If there is none, disconnect the downstream devices starting at the first jack to see if the problem is a downstream short that is pulling down the connection. With the feed from the central office disconnected and all devices unplugged, ordinary ohm readings with a multimeter are useful. You can shunt out the last jack and take it from there, taking care that the line is disconnected at the first jack.

> Short pieces of the jacket of Cat 5e cable with conductors removed make good protective sheaths for small conductors such as motor leads that may chafe. These pieces can also be used as sleeves to identify multiple ends that are to be terminated later. A fine-point felt-tip pen works for labeling the individual tubes.

All of this assumes that you know the order in which the jacks are fed, and from which end. This may not be evident if the cabling is concealed behind wall or ceiling finish. In any event, there are some procedures to follow before fishing any new lines. You can put your toner on the jack and work back toward the central office, checking upstream for dial tone and downstream for tone. This should reveal the location of the faulted line segment, if that is the case. Ohmmeter readings will let you know if it is a short or open. If you find a bad line, and if the line was run in Cat 5e so that there are unused spare pairs, you may find a good pair and swing over to it at both ends. This could be the case if an errant nail damaged one pair in the line. Otherwise, it could be necessary to run a new line.

An alternative method is to start at the central office. Equipment from different manufacturers includes various diagnostic capabilities including the ability to test the line. An on-screen report is generated that includes the resistance of the line read from the switch. Depending on the result, it may be possible to locate and repair the fault, swing over to a good unused pair, or run a new line.

If there is no dial tone at the central office where the pair comes out of the switch (with the downstream line disconnected), you probably have a dead card. It can easily be swapped out and reprogrammed if you have a good card on hand, but beware of static electricity in handling a card. There should be a grounding bracelet available at the switch enclosure. If not, consider installing one. In addition, it is important not to have bad cards floating around and mixing with the inventory of good spare cards, so keep track, label, and document everything. Cards are very expensive.

Why Use Cat 5e?

Any new lines should be run in Cat 5e. This makes a high-quality inexpensive phone line that is easy to work with, and it provides excellent color-coding with the potential to upgrade single-line phones to two lines and to replace damaged pairs without running new lines.

When bringing Cat 5e into a jack, strip back the outer jacket so that it ends approximately ½ in. inside the enclosure. Leave all pairs the full length, and wrap unused pairs around the part of the jacket that is inside the jack. This makes a good strain relief, keeps the extra pairs compact so that the jack is not overcrowded, and makes them available for future use.

Noisy lines are similar to dead lines, as far as troubleshooting techniques are concerned. Noise is caused by an intermittent short or open anywhere along the line, in the phone, or very often in the receiver, receiver cord, or line cord. These faults can often be located by wiggling the connections at both ends of the cords and at the jack, or by phone substitution. Underground extensions, such as when there is a phone in an outbuilding, are a frequent source of noise, often due to water in the conduit (usually PVC), frost movement at either end, or moisture in an unheated building. This can be checked by temporarily disconnecting the underground line at its source. It is necessary to disconnect both wires because noise will result if either one of them is subject to intermittent grounding. If there is an ac hum, it may be from the central office or, more likely, the phone line is run close to a power line, fluorescent ballast, motor, or some other nonlinear load.

Audio and Video Equipment

Stereo sound equipment enhances the listening experience by providing a realistic approximation of a live performance. The whole idea behind good sound reproduction is fidelity, which means accurate rendition of the original. This is most relevant to music. Enthusiasts go to great lengths to create as near perfect sound reproduction as possible, and while the technician may not share fully in this passion, it must be respected when working on this equipment. For a start, a quality setup is essential to get the most out of the equipment.

When finished soldering, retin the soldering iron tip so that it is ready for next time. Tinning the tip prevents corrosion. If the tip is corroded, the oxide coating acts as insulation and prevents heat transfer so that it is impossible to solder. Disconnect the soldering iron and apply solder as it cools. To clean a corroded tip, heat it up and wipe it on a damp sponge. Apply flux and solder as it cools down.

The fundamental idea behind stereo is that at the time of the original recording, two (or more) microphones are placed, let us say, at opposite sides of a stage or, for an orchestral performance, at different locations among the musical instruments. Recording engineers make all of this into a high art form, with many tweaks and adjustments that make up the final product, but the basic mechanism is simply that two separate signals are recorded and reproduced so that they can be played back as two channels. For years, this was done predominantly on vinyl LPs, then cassette tapes, and currently on the wonderfully convenient and durable CDs. (A recent development is MP3, an encoding format for digital audio that permits compression by reducing the audio resolution that is beyond what most people are able to hear.) The two sound tracks, at the listener's location, are separately amplified and the two signals are sent to two different speakers, making for a superb listening experience where the spatial characteristics of the original concert hall or recording studio are recreated.

Optimum Sound

Setting up sound reproduction equipment, at the most advanced level, is a job for professionals with specialized knowledge and expertise in the area. There are, however, a few basic principles that should be observed to get good sound.

First, the speakers must be in phase. The two speakers should simultaneously make compression waves in the air that reach the listener at the same time, as opposed to one speaker pulling away as the other pushes in. Phase may be changed by reversing the wires of just one (not both) of the speakers. It is usually done by trial and error, to see which sounds better. Some advanced stereo systems have incorporated a phase switch. You can make one of these using a four-way switch in line with one of the speakers. When the stereo outputs are out of phase, the sound will appear muddy and the stereo effect is indistinct, so the difference is immediately evident. One warning: Do not change phase, either by exchanging speaker wires or by using a phase switch, with the amplifier powered up. In fact, an amplifier should never be run without speakers or a dummy load, as this will cause the output voltage, being open circuited, to rise to a dangerous level and destroy the output transistors.

In a rectangular room, speakers should be placed equidistant from the corners along the long dimension. They should toe in a slight amount but not excessively, and should be out from the wall a little. They should also be elevated above the floor. The room should not be heavily upholstered and composed entirely of acoustically absorbent material, but neither should it be all hard surfaces—rather, somewhere in between. Either of the extremes will make for an artificial or overly stylized sound. The couch or seating for listeners should not be backed against the wall, but should set out slightly to minimize the effect of inappropriate reflections. These design considerations for a stereo installation will affect the quality of the listening experience. A few minor adjustments along with correct phasing will be very noticeable to the aficionado.

As for stereo troubleshooting and repair, the equipment is easy to diagnose. This is because with two channels, if one is outputting a bad signal, it is possible to take electrical measurements at identical points within the circuitry of each channel and compare the readings. This technique applies to resistance readings, voltage readings, voltage drops across resistors, capacitor tests, etc.

A stereo amplifier is easy to fix. If the unit is dead, check for a cord, switch, or power supply problem. If one channel is out, but not the other, take readings at various points and see if there is the same reading in the other channel. Suspect a bad speaker, which can be checked by applying voltage from a dry cell. As polarity is reversed, the cone should either pull in or push out. This technique can be used to determine correct phasing in a stereo system.

If the whole unit is dead, verify that there is branch circuit power, and then check the cord. Like all appliances, a frequent fault is in the power cord where it enters the cabinet. Next, check transformer input and output ac voltages. Shorted windings can make for bad output voltages. If they become too high due to shorted primary windings, which would change the turns ratio, downstream components can be damaged. More likely, however, a winding is simply open or grounded out. Resistance measurements are useful for checking the transformer windings, but just make sure the unit is unplugged and power capacitors are discharged through a load. A bad power supply transformer may have visual signs of overheating. Power supply diodes and capacitors can be checked as well as any in-line fuses, which may or may not appear visually blown.

By its nature, the power supply is intimately connected at various points with every stage in the amplifier, so if there is an excessive power supply output voltage, there can be open or shorted components downstream. Visual inspection may reveal this damage. If the power supply is good but neither channel is operating, look for a bad common ground return.

If one channel is working, but not the other, compare electrical values at identical points on both sides. It is good to start at the speakers because the biggest power flow is here and therefore there is a great chance for mischief. There are several ways to check speakers— exchange them between channels, take out of circuit ohmmeter readings at the terminals, or connect a dry cell to the speaker inputs and observe whether the cone moves. (Similar tests can be used to check the speaker wires.) Very often there is an open coil or, especially for large woofers, a broken lead that is too short to solder. Many speakers have impedance matching transformers mounted on them. They go bad when called upon to handle too much power, particularly when the volume is turned up high. Replacement speakers often come without transformers, so when installing a new speaker it is necessary to take the transformer off the old speaker and mount it on the new one. This involves drilling holes in the new speaker's metal frame. Use a magnet to catch the steel particles. While we are on the subject, it is worth mentioning that a large discarded speaker is an excellent source for a very powerful permanent magnet, which is useful around the shop or on the job for such diverse tasks as retrieving dropped tools and hardware that have become inaccessible, cleaning up spilled nails, holding steel objects in place on the bench to aid in difficult soldering jobs, and setting off fire alarm heads in order to test them for scheduled maintenance.

High-end stereo equipment is very expensive, so great care must be taken when working on it. If an internal fuse has blown, it may be because the fuse element has become fatigued and its time has come. It is also possible that a line surge has occurred due to nearby lightning or inductive kick from a load being interrupted quickly. In these situations, a fuse can be replaced and the equipment fired up. However, there are instances where a blown fuse indicates a problem within the equipment such as a shorted lead, wiring fault, or component failure that caused the circuit to draw abnormally heavy current. If so, a replacement fuse should cut out before further damage occurs. However, this is not always the case. Therefore, when there is a blown fuse, especially in expensive equipment, the sensible course of action is to trace the circuit downstream from the fuse. Start with a careful visual inspection. If it is a printed circuit, see if there is a partial solder bridge between any two adjacent traces. In addition, some sort of conductive debris or metal filings could be the problem. A minute fracture in the circuit board can fill with dirt that may suddenly become conductive if there is the slightest amount of moisture. These problems become more acute at higher voltages. Ohm measurements are helpful.

Alternatively, capacitors and resistors often appear discolored or distorted when they go bad. Test every downstream component before replacing the fuse. Above all, avoid the impulse to go to a higher ampere value for the fuse. This can endanger a single component or the piece of equipment, or even cause an electrical fire within the building.

The Code permits ungrounded systems in certain applications. Of course, even an ungrounded system has to have a green or bare equipment-grounding conductor connected to a ground electrode system complying with the full set of Code rules.

In troubleshooting stereo equipment beyond the power supply, a useful technique is to insert a signal into the front end. A CD output will do, but the ideal is a uniform audio tone, the same level being fed into both channels. A signal generator is excellent for this purpose, and some technicians are fortunate enough to have one of these in addition to an oscilloscope. For many, this is more an aspiration than a reality, and it is necessary to work with what is available or what can be fabricated. There are audio oscillators available for download from the Internet, and it is possible to put together a computer-based oscilloscope, especially if you have a laptop you are willing to dedicate for the purpose. However, this is not equivalent to having top of the line equipment, albeit much less expensive.

You can make dummy loads out of 4- and 8-ohm power resistors, so that an amplifier being worked on will not overload speakers that might be more costly than the amplifier. A bench power supply is great to have, but alternately you can put together something from the power supply of a discarded amplifier. The plan is to power up the amplifier gradually, so that if there is a short it will not come on full force.

Audio Fundamentals

If an amplifier had been working properly but abruptly developed a fault, it may suffer from a component that has experienced spontaneous premature failure, solely due to a manufacturing flaw. Semiconductors are prone to this syndrome, but it happens to capacitors also. Another cause of equipment failure is a poor solder joint. Insufficient heat or a bit of oxide during manufacture can result in a solder joint that works for a while because of pressure contact, but eventually fails because alloying of the metals did not take place. A bad solder joint can be difficult to detect, either visually or by means of an ohmmeter. If all else fails, remelt every joint on the board with a pencil-tip soldering iron, using heat sinks where needed to avoid damage to sensitive components, especially semiconductors. In the same spirit, pull apart and reconnect all ribbon connectors and slide-on terminations as this can repolish corroded metal surfaces.

On the subject of audio equipment, we might as well take a look at the superheterodyne radio receiver. A receiver is an amplifier (often stereo because some FM transmitters offer two-channel stereo programming) combined with a tuner and radiofrequency (RF) amplifier. These units are not normally repaired unless it is a superficial power supply or speaker problem, although some high-end receivers are valuable enough to justify a full-scale tear down and extensive parts replacement. Many receivers are combined with a stereo CD player, graphic equalizer, alphanumeric display, and other features. These units offer advantages in troubleshooting, if not repair. If the CD player produces beautiful sound but the receiver is silent, you can immediately eliminate large amounts of circuitry as being at fault.

Even if you do not foresee a situation where you will be servicing radios or TVs in the future, it is still instructive to know how these things work, and the knowledge will provide insight into other types of electronics.

The superheterodyne receiver was a remarkable invention with its origins in the early twentieth century, and today virtually all radios and TVs operate in this way. With separate

RF, intermediate frequency (IF), and audio frequency (AF) amplification, these radios are highly amenable to the diagnostic procedures that we have been discussing.

A radio antenna produces an output in the microvolt range. This RF signal, after tuning by means of a resonant circuit and detection by diode action, is barely audible without amplification through headphones. At the transmitter, the RF carrier wave is modulated by an audio signal. The process may be amplitude modulation, frequency modulation, or phase modulation. At the receiver, this modulated signal is amplified in successive stages, the output of the first stage becoming the input of the second, and so on. Then, through a process of detection and filtering, which resembles rectification in a power circuit, the RF component is removed, leaving an AF that is intelligible to humans. That signal, in turn, is amplified until it is strong enough to drive a speaker. This generic radio worked after a fashion, but had some severe drawbacks. For one thing, many stages of amplification were required, which translated into the need for many tubes, at that time quite expensive. Moreover, the signal was not very stable; whistles and noise accompanied the final product.

About the time of World War I and in response to allied military needs, a new product was developed, known eventually as the superheterodyne. It used a technique of mixing frequencies to produce a more usable resultant. If two musical tones, ideally pure sine waves, are present in the same air space, they will mix, their own identities intact but with a portion of the energy going into two additional tones. These frequencies are equal to the difference and sum of the original frequencies. The sound made by the difference of two tones that are close together is a very noticeable low-frequency "beat," which you have probably heard. The same thing happens at much higher frequencies, electrically, when radio frequencies are mixed. The resulting IF, still modulated, is more amenable to amplification for several reasons, as we shall see.

The earliest arrangement was for the transmitter to broadcast both the modulated signal and an oscillator-generated RF signal. The two could be combined to produce the desired IF. Soon a superior protocol emerged whereby the required frequency was created locally, meaning within each receiver. This was accomplished easily—an oscillator is just an amplifier that is biased so that its output becomes a pure sine wave. You may have noticed that radios with knob-controlled variable capacitors have two of the components attached to a single shaft. One of them tunes the modulated RF while the other varies the locally generated oscillator output. When the two combine, a single unchanging IF is produced, and it is this signal that is amplified prior to demodulation, or detection as it is known. The advantages of a single IF as opposed to the varying RF for amplification are twofold. For one thing, the frequency is lower than the RF as broadcast by the transmitter, so there is not as great a battle against unwanted capacitance and attendant losses. This means fewer stages are needed. Second, since there is a single IF rather than the varying RF, resonant frequency filtering can be accomplished by a single set of components rather than one for each broadcast channel. Because of these inherent advantages, superheterodyne technology remains the dominant methodology at present, even in video reception.

What does this mean for troubleshooting? A signal can be injected at the input of the first IF stage and voltages or oscilloscope readings can be checked at the output of each stage. Some schematics show an oscilloscope graphic at various test points, and this facilitates locating the circuit and component that is at fault. At IF frequencies, the process is rather more manageable than in the varying high-frequency regions upstream.

When wiring a three-phase motor, it is important to balance the phases. The idea is that one leg of the supply will likely have a slightly higher voltage than the others do, and one leg of the load may draw a little more current than the others do. Hook up the motor with the desired rotation and then take clamp-on ammeter readings. Then roll the connections without reversing any two legs, which would reverse the rotation. Take readings for all three configurations and whichever set of readings is most uniform is the right hookup. After connections are finished in the motor terminal box, the wiring has been stuffed in, and the cover is screwed on, go back to an upstream access point such as the three-pole breaker in the panel, and with power off take ohm readings. The resistance in all three phases should be substantially equal, say 3.9 ohms, and there should be no resistance to ground at a high megohm range setting. That means wiring is correct, except that rotation may still be wrong. Often it is safe to try out the motor briefly to check rotation, but beware of some pumps, which can be instantly damaged (seal ruined) when run the wrong way.

Next on the agenda is the subject of video, which is several orders of magnitude more complex than simple amplifiers or radio receivers. This is especially true where color reception is concerned. While it is beyond the scope or intent of this book to provide the reader with any significant knowledge or expertise in color TV servicing, we can describe some of the basic principles and indicate some resources that would be helpful, if that were your inclination. To become adept at this trade, a rather thorough knowledge of electronic theory is required, plus a large amount of TV-specific learning, and perhaps several years' experience, preferably working with an experienced technician who is willing to share information.

Electronics courses with a TV servicing slant vary widely in quality and cost, and the two appear to have little correlation. Of course, there is a lot of free information on the Internet. If you have a specific question regarding a project you are working on, type the make and model into a search engine and put your concern in the form of a question. Most likely, you will get a focused answer.

TV Repair Procedures

Television, as mentioned, is fairly complex. For a start, there are potentially lethal voltages that supply the picture tube, and anytime you go inside the cabinet, you have to be aware of that fact and protect yourself accordingly. It has to be emphasized that certain components and wiring remain energized at much higher levels than the 120/240 volt system that supplies the TV, even when it is turned off and unplugged. This is because capacitors, including the picture tube, which can function as a capacitor, have the ability to store a very high charge for a long time. Here are some methods of protection:

- At the bench, place a large, thick, dry rubber mat on the floor, so that when you work on equipment you are not grounded. You can still get a shock, but it will not be as severe. Your bench should be made of insulating material, that is, not metal, and there should be no grounded surfaces that can be inadvertently contacted, such as grounded metal light fixtures.

- Be aware of which parts within a TV carry high voltages. A clear indication is thick insulation and heavy rubber boots where the wires terminate. Do not touch them, even if you think they are de-energized. Use insulated tools, not your bare hands.

- Discharge all large electrolytic capacitors and high-voltage wiring by shunting with a power resistor that is known to be good. After you think the voltage is down, shunt them out with an insulated copper wire, just to be sure. Do not forget that this process has to be repeated each time the equipment is fired up.

- If it is necessary to take meter readings on a high-voltage circuit, de-energize and discharge everything, and then connect the meter using alligator clips, as opposed to touching the live terminals with hand-held probes—that is a recipe for disaster.

- If you get a significant electrical shock, seek medical attention. An individual can be shocked and feel fine afterward, only to expire several hours later due to heart damage.

These cautionary notes apply to all electrical work, but especially for television servicing because of the presence of the high voltage necessary to deflect the beam of electrons within the picture tube. Generally, the bigger the picture tube, the higher the voltage.

The bench should be spacious and uncluttered, if you are to work safely and efficiently. When the chassis is removed from the cabinet, the power cord normally becomes disconnected. Technicians who use a cheater cord to override this safety feature should be knowledgeable about the hazards involved and capable of protecting themselves and others from contact with high voltage. Do not leave a live, undischarged chassis unattended. A bench-mounted mirror is helpful so that you can work at the back of the set while observing the screen.

Some problems are temperature dependent. If a fault consistently surfaces some time after the TV is powered up, it is possible that a component has a temperature-dependent fault, such as a minute crack that opens up. To locate the faulty component, Cold Spray or Circuit Chiller works quite well. After the fault appears, direct the spray very briefly at suspect components, one at a time, to see when the fault goes away. This is not a fix, just a diagnostic technique. Similarly, a heat gun with a small metal funnel attached to direct the airflow can bring out a quiescent defect.

On the wall above your bench, you can mount a 4×4 surface box with two receptacles wired in series, preferably on a dedicated circuit. A plug-in bulb socket with an appliance bulb can be inserted into one of these receptacles and it will serve to limit the current available at the other receptacle, protecting equipment that has a dead short. The lower the bulb wattage, the more the current is limited, that is, the greater the protection. An appliance bulb is a good choice for a start. Also in series, at a convenient location so you can hit it quickly, install a single-pole toggle switch, or a normally open push button.

An isolation transformer is valuable. Remember that grounding does not pass through a transformer, so use of this device will temporarily remove the ground connection from a TV chassis. You can fabricate an isolation transformer by connecting the secondaries of two identical power transformers from discarded TVs. Choose small TVs that have transformers with clearly marked ratings so that you do not get involved with excessively high voltages, and carefully insulate the wiring.

Computer Troubleshooting and Repair

The familiar desktop or laptop computer incorporates two distinct areas—software and hardware. Software is a broad and somewhat diverse term, so to begin on secure footing we will consider hardware first. We all know what that means. Hardware is made up of the physical components including the central processing unit (CPU) plus all peripherals such as printer, keyboard, monitor, speakers, and so on. Also included are cabling, satellite dish if used for Internet connection, and so on. What these all have in common is a physical reality. A compact disc (CD) is a piece of hardware, while the information it contains is within the category of software, whether or not it is used for programming.

When it comes to computers, everything falls into one of these two categories. When a problem arises, it is generally useful to ask, "Is it a hardware or a software problem?" Since the two are interrelated, it is possible for the damage to be in both areas simultaneously, and indeed these problems may feed off one another.

Of course, there are instances where it is an obvious hardware problem. When the computer does not power up at all, begin by checking the branch circuit supply and the power cord. If they are not at fault, go on to measure voltages at the power supply, input and output. If there is a blown fuse, proceed to check the downstream components carefully before throwing in another fuse.

In addition, it often happens that the cooling fan stops working. (There may be more than one fan.) The fan is usually audible from the outside. If it is making an unusually loud or altered sound or no sound at all, power down the computer, check the supply voltage to the fan, and if the fan is at fault, replace it. Some computers shut down in response to a perceived temperature rise while others continue to run, rapidly frying valuable microchips. Fans are very inexpensive, so do not bother trying to oil the bearings.

BIOS Start-Up

If the CPU power light goes on, but the monitor remains dark, it could be strictly a monitor problem. It is also possible that the monitor is good, but the computer is not telling it what to do. This is where the basic input/output system (BIOS) comes in.

As most schoolchildren understand these days, a computer contains two separate types of memory. Primary memory refers to the systems that operate more or less instantaneously, this being within microchips. This random access memory (RAM) is further subdivided into volatile and nonvolatile memory. Nonvolatile memory is permanently wired into semiconductors, while volatile memory is available for both computer access and manipulation.

The other broad category of computer memory is the hard drive. This is a mechanical device in which information is recorded, retrieved, rearranged, and erased as needed. It is much slower but has vastly more capacity than the semiconductor-based memory. It is instructive to take a discarded computer apart, see how the hard drive is constructed, and how it is cabled into the logic board and, ultimately, the outside world.

When a computer is first turned on, the BIOS function comes into play. This is a type of firmware that is built into the computer in the form of a specialized nonvolatile microchip located on the motherboard. There are different BIOS chips for each type of computer, and they are designed to wake up the computer before the hard drive enters the picture.

BIOS performs several functions, typically configuring hardware, setting the system clock, enabling or disabling system components, selecting devices for booting, and initiating password prompts. This means that if the monitor fails to come up, it could be a hardware problem in the monitor itself, in the CPU BIOS chip, or in associated circuitry.

Computer monitors are a bit more complex than TVs as far as servicing is concerned. They are similar, but computer monitors have auto-scan, high scan-rate deflection electronics, and more elaborate power supplies. An auto-scan monitor automatically determines the input scan rate and chooses the correct horizontal and vertical deflection and power supply voltages as needed.

As for monitor repair, beware of high voltages. Like a TV, a computer monitor retains for a long time a large amount of electrical energy even when turned off and disconnected from the power source. Please review the cautionary material in Chap. 7 for information on protecting yourself from this hazard.

When installing a submersible well pump of 1 horsepower or less, you can dispense with the pull rope. It is just in the way and the pipe-wire combination is capable of pulling up the pump if that ever has to be done. Some installers leave out the torque arrestor, but that is a bad practice because reverse torque when the motor starts and stops can lead to a chafed wire and ground out one of the conductors.

Used monitors are abundant because outdated or nonfunctional computers are frequently discarded along with good monitors. Check with large office buildings and businesses. There is high interchangeability for monitors if the cabling is compatible; so, many times an extensive monitor repair procedure is not warranted. However, if you are interested, here are some monitor symptoms with remedies:

- If color or brightness is seen to be variable, or if there are problems with size or position, look for bad connections. They may be either inside the monitor or at cabling terminations, including inside the computer.

- If ghosts, streaks, or shadows are observed, check the input signal. They may also be caused by bad or excessively long cables, connections to other monitors, or other types of cabling design or installation flaws.

- Electromagnetic interference (EMI) or line interference is caused by nearby motors, nonlinear loads, or bad utility power. The solution is to relocate equipment or wiring that is problematic, or construct grounded shielding.

- Reduced brightness is a chronic problem for a monitor as it ages. Internal or external adjustments often but not always provide a solution.

Sometimes, ascertaining whether a fault is in the software or hardware can be difficult. If intermittent freezing takes place, it could be due to overheating, a hardware fault. Look to the power supply or the fan.

Alternatively, it could be due to a software failure, and there may be no immediate way to determine where the fault lies. However, if you keep thinking about it and performing standard troubleshooting procedures, you should be able to come to an understanding of the machine status, and find whether it is capable of being repaired.

If there are intermittent freeze-ups that require rebooting, notice whether they take place when the temperature in the room is excessively high. Try placing a small electric heater near the computer and see if that brings out the malfunction. If so, try using the heat gun very gently on various components to locate an intermittent defect.

Hardware versus Software

If, despite your best efforts, a hardware-based fault does not seem to emerge, you will need to take a second look at the software side of the picture. Recall the history of the malfunction, conferring with users of the computer. Did the problem arise immediately after a new program was installed? Some programs in conjunction with other programs give rise to conflicts, and they may cause the computer to crash. Use a half-splitting troubleshooting technique to localize the application that is causing the problem. You can temporarily disable programs to see which one is responsible. If one of the two that is causing the conflict is expendable, that is the end of the story. If not, see if there is a newer version available. Otherwise, simply removing it and reinstalling it may rectify the situation.

If this does not work, try a PC or Mac repair program, found under utilities or a similar menu entry. Do not be afraid to defragment the hard drive, as this should be done from time to time and may solve chronic memory shortages.

If all else fails, it is possible that the operating system has lost data or acquired one or more corruptions. Use the restore disc that came with the computer.

As a diagnostic procedure, safe mode is useful. For a windows-type computer, it is instituted by holding down the F8 key as the system boots up. For a Mac, hold down the Shift key while booting. This results in a start-up with no extensions, and is useful if a conflict is suspected. For a Mac, begin by shutting the computer down. Then press the power button. After but not before you hear the start-up tone, depress the Shift key. When the gray apple icon appears, release the Shift key. To come out of safe mode, restart the computer in the normal fashion, without holding down the Space Bar.

When repairing an amplifier or similar equipment and you cannot find the problem, try going over all the solder joints with a hot soldering iron. With time, one or more of them may have oxidized at a cold joint making a high-resistance connection. Another possible remedy is to separate and reconnect all the ribbon connectors, which polishes and restores the contacts.

Laptops constitute a special case of computer repair, not because the operating system or electronics is appreciably different, with the exception of the battery and battery charger, but because opening up the laptop and access to sections within are more acute, due to efforts by the manufacturers to make them slimmer and more compact. This impulse is, in my view, misguided. Who cares if the laptop weighs a few ounces less or is a fraction of an inch thinner? It is very easy to carry around anyway unless you are planning to take it with you on an ascent of the Mount Everest.

The MacBook Pro with Retina Display carries this trend to such an extent that, although it is a beautiful human artifact, it is virtually unrepairable outside of a specialized shop. It takes special tools and procedures to open it up. To change the hard drive or battery is difficult beyond description. (The battery is cemented in place!)

In contrast, the previous MacBook Pro 15 in. is robust with bountiful memory and advanced features, yet battery, trackpad, RAM, hard drive, optical drive, airport card, magsafe board, and fan are all easily accessible.

Apple introduced the MacBook Pro in 2006. Externally, it resembles its predecessor, the Powerbook G4 Apple laptop. The changes to the casing are minor, but inside it is a completely new computer with a large capacity hard drive and a lot of RAM.

When working on this computer, as with any sensitive electronic equipment, take whatever measures are possible to guard against static electricity damage to semiconductors. An antistatic bracelet attached to a grounding terminal will work. If this is not available, touch a grounded metal surface periodically in the course of your work. Hold printed circuit boards at the edges and keep replacement semiconductors in their original conductive foam packing, which keeps the leads shunted together so that they cannot be at different voltage levels.

As with any laptop, during a repair you want to remove the battery at the earliest opportunity. For this model, there is nothing to it. Just turn the laptop over, and rest it on a clean cushioned surface. There are two latches at the bottom of the case near the middle. Slide them toward the back, and then the battery will pop out. After the battery has been removed, five screws are visible. Of these, three screws with large heads hold the memory bay cover in place. Remove all of these screws, noting where each type goes. Take copious notes at each step of the teardown process, so that there are no questions during re-assembly. Note that three of the screws are shorter than the others.

After the screws are removed, take off the memory bay cover by sliding it toward the front of the computer. Once the memory bay cover has been removed, you will see two RAM slots. They will contain whatever memory was included when the laptop was originally purchased or installed in the course of an upgrade. There will be either one or two memory modules, held in place by metal clips. Move them out of the way and slide the modules out.

Reverse the process to reinstall the memory modules or replace them with new modules. The maximum upgrade involves two 1-GB memory modules for a total of 2 GB of RAM—that is a lot of memory.

The last article of the National Electric Code covers Network-Powered Broadband Communications Systems. "Network-powered" means that in addition to a data signal for telephone, cable TV, or interactive services, the line also carries a dc or ac voltage which is used to power any electronic components such as required for amplification along the way. In the early years of telephone technology, each phone contained its own batteries. If you look inside a modern phone, you will see solid-state components for amplification, but no power supply or rectifier. A dc voltage is provided by the utility and accompanies the voice signal on a single pair of conductors.

Next on the agenda is to remove the top case. There are 10 screws altogether. After they are removed (using a Torx TR tool), pull out the top case enough to reveal the cable that connects it to the main logic board. Sometimes the top case is reluctant to separate at first. The perfect tool to aid in this is a plastic guitar pick, which will slide in gently and not mar the edges. Disconnect the top case. Now the hard drive is accessible. If the object of this repair is to replace the hard drive, disconnect the hard drive cable from the main logic board. Pull the cable evenly and it will come free. On the right side are two screws holding the hard drive bracket. Remove the Bluetooth module and the bracket. Everything comes apart easily. Remove the hard drive and disconnect the main interface cable.

To install a new hard drive, simply reverse the steps above. That is it, as far as the hardware is concerned. It will still be necessary to format the new disc and install the operating system. Directions for doing this are on the Internet, and are not overly difficult.

More Mac Subsystems

Going deeper into the MacBook Pro, it is feasible to replace the Bluetooth module, the Superdrive, Airport Express including antenna cables, keyboard, speaker assembly, Magsafe board, logic board, display, inverter board, and other components.

General troubleshooting techniques apply to both hardware and software problems. Hardware defects are easier to pin down, but the parts may be more expensive. Software troubleshooting requires a measure of expertise, but the overhead is often zero.

In looking for the source of hardware problems, an important clue is provided when you think back (or discover from the user interview) to determine if a specific event initiated it. Did it first occur when a new cable or peripheral device was installed? If so, check the integrity of connections.

If a computer boots up successfully, but a hardware device or software program fails to perform, there is reason to suspect that it is a driver problem. A driver is a small application that controls a device, mediating between it and the operating system. All devices have drivers. Devices, in this context, are units of hardware including printers, keyboards, disc drives, etc. The computer's operating system contains the driver for the keyboard, but other devices need to have their specific drivers installed in the computer. A printer, to take one example, usually comes with a CD containing the driver. If the CD is not available, the driver can be obtained as a free download from the Internet. Type the make and model of the device, the computer platform (Mac or PC), and "download driver" into the search engine.

Very frequently, a device will exhibit erratic behavior or cause the computer to crash. By downloading a new driver, you can often resolve the problem.

There is a very powerful tool for diagnosis of computer problems. You have undoubtedly noticed that when a computer starts up, there is a characteristic chime or tone. It is initiated by the BIOS application and indicates that diagnostic tests have not detected hardware or software problems. The PC terminology is POST. This means Power-On Self Test. Depending upon the computer and type of BIOS chip that it contains, the error codes will vary, as will the amount of information available to aid in troubleshooting. As an example, the following is a list of the standard original IBM POST error codes:

- A single short tone: System is good.
- Two short tones: Error is shown on screen.
- No tone: System board or power supply malfunction.
- Continuous tone: Keyboard, power supply or system board malfunction.
- One long and one short tone: System board failure.
- One long and two short tones: Display adapter malfunction.
- One long and three short tones: Enhanced graphics adapter malfunction.
- Three long tones: Keyboard card failure.

These error codes can take you directly to the problem. To decode tones (or flashing lights), type the computer make and model along with the appropriate query into the search engine and you will find the answer.

CATV cable typically has an ac voltage of approximately 60 volts, which can be introduced at any point along the line. The amplifiers, which may be every 400 ft along the line, contain transformers and rectifiers, which power the solid-state amplifiers. The ac is blocked so that it does not reach the TV.

A common reason for a TV cable system to go down is loss of ac power at the service, and the repair can be as simple as resetting a breaker. An important troubleshooting technique is to take voltage measurements at the input and output of each amplifier along the line. These amplifiers are in weatherproof gasketed boxes, which can be opened to change modules. Also, look for problems with the connectors.

Most electricians and electronic technicians are computer literate. They have no problem configuring Internet connections, working with advanced preferences, going into the MS DOS or command line mode to ping an external device, or performing other troubleshooting tasks. Any of these things should be easy for you and if you do not know how to do them, you do know how to go into an Internet information-gathering mode and get on with the task in short order.

However, programming is a different matter altogether. If you have looked into this field at all, you may have been overwhelmed and come to regard it as akin to learning a foreign language, the only difference being that everything has to be letter perfect just to be acceptable.

Excellent online programming courses, free of charge, may be found at www.codeacademy.com. It is a remarkable interactive site that will have you actually writing lines of programming code immediately.

A good, user-friendly entry into computer programming is learning to write hypertext markup language (HTML). It is easy and you can get results right away. HTML is used to compose web pages. You can make a complete 50-page website right on your computer and, if you have a browser such as Internet Explorer or Safari, you can display the finished product complete with color graphics, sound, video, and so on. You can compose any number of websites and save them on your hard drive, even if you do not have an Internet connection.

There are quite a number of web-authoring programs available. Some are free and others are expensive. The idea with these programs is that you can compose a web page using an interface that displays the finished product as it will appear online. Then, the program converts your work into HTML and with a click of the mouse the site will be uploaded to the web host you have chosen and instantly your work is published on the web.

HTML Coding

I have tried some of these web-authoring programs and find that they are more bother than they are worth. It is easier to simply use HTML coding and make the website that way. You do not need to learn the HTML tags; just refer to a chart as needed. Soon you will learn the tags you use most, and the process of making a web page will become logical and spontaneous.

To compose a single web page, use the text editing program that came with your computer. For Mac it is TextEdit and for PC it is NotePad or some successor. Open a new document in that program and start typing the HTML for the web page you want to make.

Anytime you connect to the Internet and, via your browser, you have a web page on screen, it is possible to view the HTML that created that page. Click on VIEW SOURCE or DOCUMENT SOURCE, and the HTML for that page will appear on your screen. This is a great way to get a feel for how HTML coding corresponds to a finished web page. You can copy and paste part or all of these documents into your text editor, change wording, colors, graphics, fonts, page layouts, etc. and customize your own web pages in no time.

> A useful item in an electrician's toolbox is the neon test light. It is good for taking readings to determine presence or absence of voltage when the exact number is known anyway. You can determine whether a conductor is grounded or ungrounded where the system is no greater than house current by putting one probe on the conductor in question and touching the other probe to the grounded chassis. The neon bulb will glow if you have the hot wire. The resistance of the bulb and series resistor, which is inside, is so high that a negligible current flows to ground.

Make a folder with all the text editor documents and graphics pertaining to each website that you create. That way, if the web host crashes and loses your data, you will not have to start all over.

Double click on the icon for one of your text editor documents, finished or in progress, and it will be displayed on the screen. Right click (for Mac, Control click) on that icon and a contextual drop-down window will appear. One of the choices will be Open With. . . . Follow that alternative and there will be a list of applications. Click on your preferred browser and the finished or in-progress web page will be displayed just as it would appear on the web. You can put that and the HTML document on screen side by side to see how changes you make in the HTML will affect the web page. Every time you make a change in the HTML, it has to be saved before it can be viewed in the browser as a web page. All of this can be done without publishing any of it on the web, and indeed without even having an Internet connection.

Let us get started. All HTML tags are enclosed in angle brackets: <HTML>. The angle brackets tell the browser that what is inside the brackets is an HTML tag. Without the angle brackets, nothing is going to happen.

There are two kinds of tags—container tags and open tags. A container tag has to be initiated and, following the material that is affected by the tag, it has to be closed. The characters that are within the two pairs of angle brackets comprise the tag. The opening tag initiates the HTML action and the closing tag concludes it. The character that denotes a closing tag is a forward slash. For example, <html> starts an HTML document and </html> closes it. Everything between these two tags is recognized by all browsers as the source code for a web page. If it is not opened and closed in a letter-perfect manner, the web page will not be displayed properly. All container tags have to be opened and closed.

There can be any number of tags within tags. Be sure to close them in the reverse order in which they were opened.

> For troubleshooting receptacles, use an appliance bulb screwed into a plug-in socket. This is easier than using a meter, where you never know if the test leads are making contact. If the receptacle is within sight of the panel, leave it plugged in while you locate the breaker. If it is not in sight, use a radio turned up loud.

The other type of tag is known as an open tag. It does not have to be closed—it is just a one-time entry. An example of this type of tag is <hr>. It directs the browser to make a horizontal line across the width of the web page. If you used your keyboard to make a horizontal line across the text editor document, it would not display properly in the web page. Similarly, you could hit the return key on your keyboard as many times as you wanted and it would have no effect on the web page. Instead, use this tag: <br>. It stands for break. In addition, you cannot make a paragraph break by pressing the TAB key as you would in a word processing program. This would appear in the text editor but not on the web page, where a continuous paragraph would be seen. To make a paragraph, use this tag: <p>. It is a container tag, but by convention the closing tag may be omitted.

A great many tags are recognized by all browsers. Each browser, however, may display them somewhat differently. Serious web composers have the major browsers in their computers and view a proposed web page in each of them before publishing it.

Principal HTML Tags

Here are some of the principal HTML tags. You can compose a sample web page using some or all of them and see how it looks.

<html> Begin html document.
<head> Begin heading.

<title> Begin title. It appears in the web page title bar and when the page is bookmarked.

<body> Begin web page material.

<h1> Largest type for heading </h1> (<h2> through <h6> are the tags for successively smaller headings).

<p> Begin new paragraph. (Closing is optional.)

</body> Close body.

</html> Close html document.

 Break in line.

<center>Begin centered material.

<hr> Horizontal line.

<strong> Bold text.

<em> Italics.

<a href="http://www.electriciansparadise.com"> Outside link.

</a> Hot link.

<a href="mailto: email@extension">Email@extension</a> Email link.

<blockquote>Insert material to be blocked between these tags</blockquote> Indent for long quotes.

<img src="picture.jpg" align=left> Insert graphics.

<table> Open table.

<tr> Open table row.

<td> Open table data.

</td> Close table data.

</tr> Close table row.

</table> Close table.

You can make an HTML template and use it every time you want to do a new page. This saves time and ensures that no items will be omitted.

If you make a website for yourself or someone else, you will need to publish it on the Internet. Web hosts, of course, have their own websites, which include a facility for checking to see if a universal resource locater (URL) is already taken. When you find an available URL that is satisfactory to you, you can pay for the service and after a short time, typically 24 hours, you can publish your site and thereafter make changes as often as you desire.

The web host will give you instructions for uploading HTML files, and its tech help will provide guidance if you experience problems.

Interestingly, HTML files do not contain graphics. They have text that refers to the graphics files, which are uploaded separately.

Typically, your homepage takes this form:

http://www.electriciansparadise.com/index.html.

Individual pages within the site take this form:

http://www.electriciansparadise.com/capacitance.html.

The final characters in each of these URLs are required to let the web host know what is going on.

You might be puzzled at the cryptic-sounding language at the top of the page, above the <HTML> tag. This is not part of the HTML. It is called the "declaration" and it is a statement of the type of HTML coding that is being used. It is optional and not necessary for the page to be displayed. However, it is believed that inclusion of the declaration makes the web page more search-engine friendly and will result in a higher search engine ranking.

Understanding Ethernet and Learning to Create a Simple Network

Network connectivity is an enormous worldwide technology, and the dominant means for achieving it on the local level is Ethernet. That is because it is a simple, reliable, and inexpensive way to wire together computers and devices so that they may be able to communicate at very high speeds. In order to do this, the problem of data collision was dealt with effectively right from the start in the 1970s, so that Ethernet quickly eclipsed its predecessor, Alohanet, and its contemporary, Token Ring.

Robert Metcalfe came up with the idea of Ethernet and, with colleagues, developed it to the point where it became the dominant technology for networking computers and devices. With some background information and a little practice, you can become proficient in creating fast and trouble-free Ethernet systems. Troubleshooting and repair are straightforward because the way in which high-speed data transfer is accomplished in this simple protocol involves the incredibly user-friendly medium known as category cable or, in its most common form, unshielded twisted pair (UTP).

Ethernet was preceded by Alohanet, which was developed at the University of Hawaii in 1970 by Norman Abramson and fellow researchers. Because this respected university consisted of a number of campuses in locations separated by bodies of water, it was necessary to employ radio transmission, so Alohanet was created to work in a wireless fashion. Two frequencies— one to transmit and one to receive—operated from a central hub and used star topology. Data was received and immediately retransmitted, and comparison revealed any data corruption that occurred. Where this kind of loss was detected, the original data was retransmitted, and this process was repeated, when necessary, until an intact message was received.

The terminal devices were teletype. With a speed of only 80 words per minute, a high network speed was not needed. New technologies were emerging in the 1970s and 1980s, however, and the demand for greater speeds required faster hardware and media (cabling) to keep up with advances. Token Ring emerged at about the same time as Ethernet. Both of them dealt with data collision, although in radically different ways. Because of its greater simplicity and ease of implementation, Ethernet finally predominated.

IBM developed Token Ring in the 1970s, and an ingenious if somewhat roundabout method of dealing with data collision was contained in this interesting protocol. A control "token" accompanied each transmission to the stations that accessed the network. Transmission was enabled by possession of the token, and those without a token would have to be patient until they received it. This method worked, but the physical substrate presented difficulties. The cabling, IBM Type 1, was a shielded twisted pair with large, failure-prone plastic connectors. The emergence of the personal computer called out for a new method of network connectivity that would address the inherent problems of Alohanet and Token Ring.

Ethernet provided the answer. Robert Metcalfe, now a member of the staff at Xerox's Palo Alto Research Center, had been engaged in the task of networking the many computers at the facility. Xerox had come up with the design for a laser printer, and the plan was to permit the computers to communicate with it as needed. Very high speed was the critical requirement.

In 1976, Metcalfe and a coworker, David Boggs, worked out the details of a new technology. They were contained in their paper titled "Ethernet: Distributed Packet Switching for Local Computer Networks."

Coaxial cable and other preexisting items were incorporated in the physical layer of this new technology. What made it work was Carrier Sense Multiple Access Collision Detection (CSMA/CD). Each access location monitors the network to determine if traffic is present. When there is a period of inactivity, any or all stations may transmit. Data collision takes place if two or more stations attempt to transmit at the same time. If this happens, they will abort and after a short period of time, retransmit. Each station waits until a random amount of time elapses before attempting to retransmit. The unique algorithm that determines the waiting time minimizes the chance for secondary collisions, and the process is repeated until there is successful communication. All of this is implemented by means of a small printed circuit at each network location. The Network Interface Card (NIC), or Ethernet card, was originally connected to the computer or device output port, but currently it is built into all computers or devices that are to be connected via Ethernet to each other, a peripheral device, or the Internet.

At first, serial bus topology was the dominant mode, but soon it was replaced by star topology. This would require more cable, but it would provide greater reliability because if there were damage to the cable at one point, only a single station would experience an outage while the rest of the network remained operational.

A major improvement in terms of ease of implementation and expense came with the introduction of UTP cable. This is known as category cable because successive versions or categories are issued with ever-increasing speed and reliability. The most widely used today is Cat 5e. (The "e" stands for enhanced). A more advanced version, Cat 6a ("a" means augmented) is available, but it is more expensive. Cat 5e is suitable for current network speeds. Eventually, as its price drops and new installation methods emerge, fiber optic will probably eclipse all UTP, but for now, Cat 5e is appropriate.

Ethernet Terminations

Ethernet cabling is terminated by means of RJ45 connectors at each end. (RJ stands for registered jack, an old AT&T terminology). The RJ45 connector resembles the RJ11, single-line, and RJ12, two-line, modular telephone connectors; only it is about twice the size. You can crimp them onto Cat 5e using the proper tool (see Fig. 9-1).

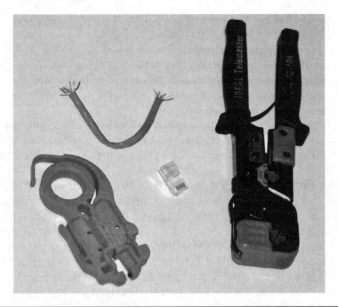

FIGURE 9-1   UTP cable stripper and RJ45 crimper.

The essence of UTP is the twisting. As noted, early networks had slow teletype machines at each end, but now enormously high speeds are expected, and this means connectivity performance will be impeded by any bottleneck, either because of the original design or because of anything that impacts the installation.

Media, the technical name for the cabling that links stations in the network, is likely to be the weak link in the chain, in terms of speed. This is because digital transmission involves waveforms that have very fast rise and fall times. The rapid changes in amplitude make for changes in inductive and capacitive reactance.

Capacitive reactance is less at higher frequencies, and where an appreciable length of cabling is involved, it appears as a shunt between the two conductors and between them and other cabling that may be nearby.

Inductive reactance is greater at higher frequencies and since it is a series phenomenon, there is greater loss for a long line that operates at a high frequency.

One way to fight off capacitive coupling is by shielding. A grounded metal barrier, placed between the conductors, will reduce signal loss of this type.

Inductance takes place when ac or pulsating dc flows through a conductor. If the current is nonpulsating dc, the magnetic field is stationary and there is no coupling to adjacent conductors. However, as frequency increases, the magnetic field forms and collapses with great energy, making for more inductive coupling and the harmful effect known as crosstalk. Moreover, there is loss of energy, making the signal weaker.

Twisting of the conductors greatly diminishes signal loss and interference. As category cable developed over the years, specifications called for a greater twist rate per inch, and this strategy was quite successful. When terminating these cables, it is important not to eliminate twists unnecessarily. Other precautions are necessary in order not to jeopardize high-speed data transmission. When passing UTP through narrow openings or drilled holes in building surfaces, leave extra wire at both ends, and do the final trimming at the time of termination, eliminating any damaged ends. If cable is laid out on the floor prior to installation, make sure that workers do not step on it. Any rough handling may disrupt the twist rate, making for reduced performance. If the cable is to be secured by means of staples or similar hardware, do not pinch the conductors as this will also impede high-speed connectivity.

UTP is available with either solid or stranded conductors. The stranded type is used for short patch cords, mostly factory-made, that will be bent repeatedly. For a permanent installation, where less flexibility is required, solid is the better choice because there is less loss at high frequencies. Whichever type is chosen, when terminating it is essential to use the right connector. Both types operate on the principle of insulation displacement, which means that you do not strip the wire, just insert it insulation and all into the opening. The stranded-type connector has sharp prongs that pierce the stranded conductor, while the prongs of the solid-type connector slide past the solid wire, making an electrical connection.

There is not enough current carried in an Ethernet cable to ignite nearby combustible material. Nevertheless, Cat 5e and other UTP cabling can be hazardous because its insulation can contribute to the spread of a fire that originated elsewhere. In addition, the insulation can contribute smoke to fire in a building. Particularly hazardous is a situation where large amounts of abandoned cable are allowed to accumulate.

National Electrical Code (NEC) has mandates designed to mitigate these hazards. For one thing, accessible abandoned cable that is not tagged for future use is to be removed. This is important from a fire safety standpoint, but who is responsible for this removal? It could be construed that if you do any data cabling installation work within a building, you are responsible for the removal of all abandoned cabling, even if left behind by a previous owner or tenant. This could be an overwhelming job. It would need to be done with great care because there is the possibility of disrupting a working network, interrupting the workflow of a large commercial organization. Questions regarding the implementation of this far-reaching requirement remain unresolved. (Cabling that is within a raceway is not considered accessible and does not have to be removed.) Interestingly, there is no such parallel requirement applicable to power and light wiring. It applies only to so-called "low-voltage" cabling.

Mitigating Fire Hazards

Another NEC requirement that is intended to mitigate fire hazards that may be present in network cabling is that the hierarchy of cable types must be observed. UTP and other low-voltage cabling are characterized by insulation type based on their ability to propagate fire and generate smoke. The cable types are:

- Plenum-rated, for use in spaces such as raised-floor systems and air-handling ducts
- Riser-rated, for vertical runs between floors, elevator shafts, and so on
- General purpose-rated, for nonplenum and nonriser applications
- Restricted-use, permitted in dwellings only

The cabling types higher on the list are less prone to fire propagation and smoke generation problems. They can be substituted for those lower on the list. Plenum cable, for example, can be used in a riser location, but riser cable cannot be used in a plenum location. The cable types higher on the list are more expensive. Nevertheless, many electricians use only plenum-type in order to simplify the inventory. Electrical inspectors like that.

Another practice that contributes to fire safety is to seal around any penetrations in firewalls, transformer rooms, equipment areas, and similar locations. Many local building codes require a 24-in. horizontal separation between wall boxes on opposite sides of an internal partition.

Before describing procedures for terminating Ethernet circuits using UTP cable, it must be emphasized that ordinary splicing techniques do not work. UTP, specifically Cat 5e, can be used to good advantage to make telephone branch extensions, anywhere from the central office to the final telephone. If it is necessary to make a splice or tap, the small blue wire nuts work fine and there is no loss of sound quality. The relatively low audio frequencies do not encumber these circuits with sufficient capacitive or inductive reactance to negatively impact performance. However, high-speed data transmission, with high-frequency components in the electrical circuits, will be impossible if conventional splicing techniques are employed.

To join Ethernet cables or split one line into two branches, it is necessary to use (in ascending order of capacity and functionality) an Ethernet hub, Ethernet switch, or router. These active devices have varying numbers of ports, so choose a device that will meet the needs of your networking project. Ethernet hubs are cord-and-plug connected to an ordinary ac supply, and they provide amplification along with multiport capability. The maximum distance an Ethernet line can go without signal degradation is generally taken to be 100 m or somewhat over 300 ft. If it is necessary to go farther, an Ethernet hub or better must be placed in-line. Of course, this presupposes an ac outlet at the location of the device, unless Power over Ethernet (PoE) can be employed.

We have seen that Ethernet lines may be used to connect computers to the Internet. (Using an Ethernet hub, two or more computers can be connected to a single satellite dish modem, and they can simultaneously download completely different files.) Computers can be networked to each other, create an Intranet, or PLCs can communicate with multiple pieces of machinery in an industrial setting, all employing the Ethernet protocol.

To terminate Ethernet lines, using RJ45 connectors at both ends, first trim back the cable so that there is not a damaged end. Various Ethernet protocols employ two or four pairs of conductors, but in either case, the ends are prepared in the same way with any unused conductors providing extra mechanical strength.

Begin by sliding on the rubber boot. It is optional, but makes for a better installation because it protects the cable from bending damage at the connector entry, and it seals out moisture and dirt.

Using a data cable stripper, trim back the outer jacket at least 2 in. Be sure not to cut into the individual conductors. Cut off the ripcord. Separate the twisted pairs and arrange them in the proper order for termination. This order will vary depending on the application, as we shall see. Make the individual conductors straight, parallel, and spaced so that they will go into the connector. Using sharp scissors or the cutter that is part of the crimping tool, make a perpendicular cut across all the wires so that they are ½ in. long measured from the jacket. Check again that all conductors are straight, parallel, and in the right order, and insert them into the connector, making sure that each wire goes into the proper hole. Using the crimper, squeeze firmly on the connector, and you are done.

If an arc-fault breaker is tripping out and you suspect the trouble is a partially severed wire inside a wall, replace the arc-fault breaker temporarily with a 15-ampere standard breaker. Use the antenna of a battery-operated radio tuned to no station with the volume high. Probe along the wall and listen for static. Start with a light load at the nearest receptacle, and go from there. When done, be sure to replace the arc-fault breaker.

Building Ethernet Cables

Before doing this job professionally, practice making a few patch cords (they are handy to have) and test them on your home computer, using an Ethernet hub. Try accessing the Internet and see if websites load up quickly.

Some individuals have tried crimping RJ45 connectors onto Cat 5e cable using pliers instead of the tool made for the task, and results have been unreliable.

There are two standard Ethernet pinouts, the T 568-A and the T 568-B.

The T 568-A pinout is:

Terminal 1 – green/white

Terminal 2 – green

Terminal 3 – orange/white

Terminal 4 – blue

Terminal 5 – blue/white

Terminal 6 – orange

Terminal 7 – brown/white

Terminal 8 – brown

The T 568-B pinout is:

Terminal 1 – orange/white

Terminal 2 – orange

Terminal 3 – green/white

Terminal 4 – blue

Terminal 5 – blue/white

Terminal 6 – green

Terminal 7 – brown/white

Terminal 8 – brown

There is a definite reason that there are two separate pinouts, and it is not just to provide an alternative. The reason stems from the fact that in each computer or other device with Ethernet output or input there is one transmit pin and one receive pin. When two terminals are networked, it is necessary to connect the transmit pin at one end to the receive pin at the other end. It would not work to connect the two transmit or the two receive pins to one another.

Computers and Ethernet hubs have opposing pin configurations as far as receive and transmit positions are concerned, so when networking them the same wire order at each end will result in correct operation. This is called "straight through" wiring. If two computers or two hubs are to be connected, the transmit and receive pins at each end must be reversed. This is called "crossover" wiring.

Referring to the listing above, to make a straight-through cable, T 568-B is used at both ends. For crossover wiring, T 568-A is used at one end and T 568-B is used at the other end. In both instances, looking down at the connector with the clip to the back and the wire entries down, the terminals are numbered one through eight, starting at the left. If a crossover cable is made, both ends should be marked with an X so that in the future it will not be used for a straight-through application.

Satellite Dishes for Communication Systems

W e have stated that the North American Power Grid is the largest and most complex machine ever built. The international system of orbiting satellites and earthbound equipment that communicates with them may not be as massive and is less extensive, but as a technological achievement it is unmatched.

For satellite transmission to work, several things have to come together. One of them is the concept of the geosynchronous orbit, and another is the parabolic reflector. Arthur C. Clarke, the great novelist and visionary, conceived of a satellite whose period of revolution about Earth would equal Earth's period of rotation—one day. If the orbit is directly over the equator, the satellite will appear to occupy the same position in the sky at all times, night and day. This is called a "geostationary" orbit. The satellite must be at an altitude of about 26,199 miles, measured from the center of the Earth. (The satellite will tend to drift a bit from its ideal location, and onboard rockets are required to tweak its position from time to time.)

What is important about the geostationary orbit for communications satellites is that it permits them to be accessed from Earth using a stationary antenna, making such communication feasible because elaborate tracking mechanisms are not required.

Another key element is the parabolic reflector. A parabola is one of the conic sections known to ancient mathematicians, and because of a unique property, it is useful in many areas of human endeavor. A parabolic mirror or dish will collect energy, such as light or sound, radiating from a distant source and bring it to a common focal point. Similarly, it will collect energy from a common focal point and radiate it in a narrow beam toward a distant point. A satellite dish used for TV reception takes advantage of its parabolic shape to gather enough energy from the distant satellite to provide a good signal. The energy is reflected to the end of the feedhorn, which is suspended by means of a mounting bracket (see Fig. 10-1).

Waveguide Technology

The feedhorn is actually a waveguide, through which the high-frequency microwave radiation that comprises the downlink from the satellite is conveyed to the low-noise block converter (LNB), which is at the downstream end of the feedhorn.

Figure 10-1 DIRECTV satellite dish with feedhorn.

This mechanism is typical of a satellite dish that is used for TV reception. Another type of satellite dish in common use provides Internet access. It is used in areas where broadband cable is not available, for those who are not satisfied with dial-up access because of its slow download speed, and for those who find the idea of linking to an orbiting satellite appealing.

> Old audio speakers are excellent sources of small, powerful, permanent magnets. Among others, they have these uses: (1) Stroke screwdriver tips several times in the same direction to keep them magnetized. (2) Use them to collect metal shavings when drilling near printed circuit boards. A metal shaving can short out adjacent traces and put a piece of equipment out of order. (3) Attach them to a piece of wire and lower into inaccessible regions to retrieve dropped tools and parts. (4) Set off smoke detectors in order to test them.

The primary difference between a TV satellite dish and an Internet access satellite dish is that the latter transmits to the satellite in addition to receiving. Many TV dish owners believe that their equipment transmits to the satellite in addition to receiving from it, and some are concerned that it could be a source of dangerous radiation. In fact, the only interaction between the user of a TV dish and the provider of the material takes place within the receiver (the cable box that is separate from the TV receiver) on a microprocessor level.

In contrast, an Internet access satellite dish transmits and receives. Transmission from the end user is necessary for the interaction that takes place as part of web browsing, email, etc. One way that you can see that a dish is capable of transmitting is to note that there are two coaxial cables that connect the modem to the dish outside—a transmit cable and a receive cable.

Never work on an Internet access dish that is powered up, other than aiming it. You can receive radiation burns if it goes into a transmission mode. In the transmit mode, the signal flow in the feedhorn and, in fact, throughout the system is in the reverse direction compared to when it is receiving. A microwave signal flows out of the feedhorn and is reflected by the dish in the form of a narrowly focused beam aimed directly at the satellite.

The satellite dish system employs several types of energy flow. This makes an interesting study, and since the functions are separated, troubleshooting is simple once you see how it all works together.

In the first place, programming originates in a studio on Earth, and is either recorded or conveyed live to the satellite. Uplink transmitters on the ground, which send programming to the geosynchronous satellite, are quite large, up to 40 ft in diameter. This makes for a strong, well-aimed signal. It is received by one of the transponders on the satellite. Each transponder is tuned to a different frequency, and each transmits the signal back to earth. The transponder's reception and transmission frequencies are different, so that they will not interfere with one another. The purpose of the satellite is to provide a single, uniform location in the sky so that fixed dishes on Earth, once aimed as part of the original installation, can remain locked on the satellite. It would not be possible for end users to receive programming directly from the earthbound transmitters because, for the most part, they would not be within line of sight. Satellites may have as many as 32 transponders, and although they are grouped aboard a single satellite, they may be thought of as autonomous units.

The microwave signal at the receiving end is reflected by the parabolic dish to its focal point, where it is picked up by the feedhorn. Then it is conveyed to the bottom end, where it enters the LNB.

The feedhorn is a waveguide, which uses a most interesting kind of technology for conveying the microwave signal, which is at far too high a frequency to be transmitted over cable. Low parallel capacitive impedance would short out the signal and high series inductive impedance would further reduce the signal. A waveguide meets the challenge by conveying high-frequency signals in a unique way. The waveguide is a hollow metal pipe having a rectangular cross section. Its operation is similar to optical fiber used for guiding the much higher-frequency wave aspect of light. The dimensions are different. The width of a waveguide is determined by the wavelength of the energy it is intended to guide. Huge naturally occurring waveguides formed by thermal layers in the ocean convey the very-low-frequency sonic voicings of whales, so that they may be heard a thousand miles away.

Ordinarily, electromagnetic energy propagates through space as an ever-expanding spherical shell, its thickness depending upon the duration of the pulse. Accordingly, its intensity diminishes with the square of the distance. In contrast, a waveguide conveys the energy along its length, confining it within the channel so that there is little loss. This is because at the inner walls of the waveguide, there is almost total reflection. The electromagnetic radiation travels in a zigzag path between opposite walls. The waveguide is necessary after the action of the parabolic reflector that concentrates the spherical wave front at the focal point because if the LNB were located there, it would make too great a shadow and block incoming radiation so that it would not reach the reflector.

Since the feedhorn has no moving parts or electronics, it does not normally require servicing unless the dish is blown off its mounting and the feedhorn becomes bent or dented.

Without loss, the microwave radiation travels to the LNB, which is firmly mounted on the dish frame. The LNB contains a printed circuit, and the signal is fed to its input from a metal rod at the downstream end of the feedhorn. This metal rod is an antenna and it conveys the signal a very short and carefully engineered distance to the LNB so that there will be no capacitive or inductive loss.

The LNB performs three major functions. It is a low-noise amplifier, block downconverter, and IF amplifier. Part of its operation employs the superheterodyne principle and, as such, it requires a local oscillator. Output is mixed with the microwave signal from the feedhorn, producing a sum and a difference frequency. The sum frequency is removed by filtering, and the difference frequency, a more manageable IF, is further amplified. At this point, it is sufficiently low to be conveyed the significant distance into the building and to the modem over coaxial cable, without unacceptable loss due to capacitive and inductive effects.

Because the LNB contains semiconductor devices, dc bias voltages are required. They are provided by the power supply of the modem, which is located inside the building. Receivers can distinguish between two signals of the same frequency but different polarization. Ordinary electromagnetic radiation is composed of energy in waves oriented in all directions. If a signal is polarized, it means that it contains wave energy organized in a single plane. In order to get more use out of available bandwidth, broadcasters polarize transmission signals. Vertical and horizontal linear polarization is used throughout most of the world, but in North America, circular polarization is used. There are two types: left-hand polarization and right-hand polarization. At one time, a probe within the waveguide was rotated by means of a servomotor in order to select polarization for the channel desired. Of course, today the same result is achieved electronically at much less expense.

In the LNB, which inputs a microwave frequency at its front end, internally generated noise is a critical factor, so the LNB has to be manufactured to tight tolerances in a controlled environment. Consequently, it is an expensive component. Like a TV tuner, it is not ordinarily repairable by a local electronic technician, and must be replaced if a fault occurs.

The signal for Internet access equipment, now at the IF level, is conveyed through coaxial cable to the satellite modem inside the building. The coaxial cable carries the receive signal in one direction and, through the same conductors, it carries the dc supply voltage in the other direction to the LNB where it is needed for semiconductor biases.

Everyone knows what coaxial cable looks like, but some of its properties are not clearly understood. For example, the dc power supply voltage is carried in the normal way as a simple dc circuit with two conductors—a hot wire (the inner pin) and a grounded neutral return (the outer shield). The high frequency signal, on the other hand, is propagated in a different manner that involves the characteristic impedance of the cable, which we shall consider in some detail.

Coaxial cable consists of an inner conductor surrounded by an insulating layer, and then an outer braided or foil shield, also conductive. The purpose of the insulating layer is not just for electrical insulation; for that purpose it could be much thinner and the cable more compact. It is intended to provide a highly controlled spacing so that capacitive coupling can be limited and precisely controlled. This permits its use as a transmission line for radio frequencies.

The inner conductor, insulating layer, and outer shield/conductor all share the same axis, hence the name "coaxial," often referred to as "coax." It was invented by an English electrical researcher and theoretician, Oliver Heaviside, and patented in 1880. Later in the nineteenth century, waveguide transmission and further technical advances were made, but the major innovations occurred in the 1930s, when closed-circuit TV coverage of the Berlin Olympics was transmitted over coaxial cable to Leipzig, and in Australia a 300-km underwater coaxial cable was laid to carry telephone and broadcast transmission.

In 1941, AT&T built a coaxial line between Minneapolis, Minnesota and Stevens Point, Wisconsin, having a capacity of 1 TV channel or 480 telephone lines.

We have seen how coaxial cable was used as an early medium for Ethernet. It was quickly replaced by UTP, which was less expensive and easier to install. Nevertheless, coaxial cable is still widely used in many applications including cable TV, video camera circuits, internal wiring in electrical equipment, high-frequency instrumentation wiring, and satellite TV and Internet access cabling.

At radio frequencies, signal transmission is in the transverse electric magnetic mode. It is carried along the cable by means of magnetic and electric fields outside of the actual conductors, as opposed to conventional electrical current that travels within the conductors.

What Is Characteristic Impedance?

For coaxial as well as UTP and other cables, an important parameter is characteristic impedance. In a TV satellite dish installation, the coaxial cable segments that run from the dish to the receiver and from the receiver to the TV both have characteristic impedance. Additionally, in an Internet access satellite dish installation, the UTP Ethernet cable from the modem to the computer also exhibits characteristic impedance.

The characteristic impedance of commonly used coaxial cable is 75 ohms. This is puzzling if you do not understand characteristic impedance. Does it relate to a specific length, so that with the conductors at the far end either shunted or not shunted, an ohmmeter reading would be 75? No. Characteristic impedance can be measured only by a time domain reflectometer, and in the ordinary course of a new installation, we will not have to get involved in that, nor in the complex calculations including conductor size, spacing, and the dielectric constant of the insulating material that are used to determine the value of the characteristic impedance.

Cable having a specific characteristic impedance is supplied by the manufacturer, and the installer generally can rely on the quality of the product. However, any mishandling including kinking or pinching or improper termination of the wire can alter the characteristic impedance, although coaxial cable is less vulnerable to these problems than other cable types.

Characteristic impedance becomes relevant when high-frequency signals are transmitted over a pair of conductors. Two conductors may be understood as made up of parallel capacitive and series inductive reactance as shown in the schematic of an idealized pair of conductors.

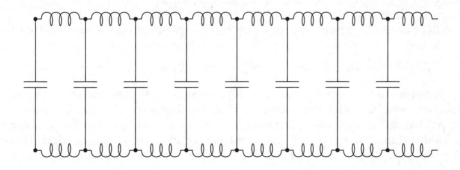

The capacitive reactance results because the two conductors are run parallel to one another, resembling the two plates of a capacitor. The inductive reactance is a consequence of the wires conducting a fluctuating current, causing the magnetic flux surrounding them to vary continually in intensity. This work requires power and so reduces the energy flow in the circuit.

If the pair of conductors in a hypothetical thought experiment is conceived as having infinite length, it will be seen that the incremental amounts of capacitive reactance will become less at higher frequencies, tending to shunt out the signal to a greater degree and increasing the current flow from the source, while the incremental amounts of inductive reactance will be greater at higher frequencies, tending to diminish the current flow. These two tendencies balance one another to make a constant impedance as the distance from the source increases, and this constitutes the characteristic impedance of the cable.

But how, you might ask, does this idealized cable of infinite length relate to anything in the real world? The fact is that a cable behaves as if its length is infinite if it terminates in a load that has the same impedance as the characteristic impedance of the cable. This is called matching impedance, and it was understood by Oliver Heaviside in the nineteenth century to be relevant to long telephone and telegraph transmission lines.

Impedances must match in source, transmission line, and load. If they do not, at high frequencies there will be reflections, collisions, and data loss. Moreover, it is necessary for the characteristic impedance to be uniform at all points along the transmission line, which is why quality control in the manufacture and installation of the cable is essential.

Characteristic impedance plays an important role whenever high frequency or long distances are involved, both inside and outside of electrical equipment. Even the traces and terminations within a printed circuit will exhibit characteristic impedance and require matching at both ends. In the repair of equipment that operates at high frequencies, it is crucial that the wiring not be altered or rerouted in a way that the characteristic impedance is changed. If the wiring is mismatched, harmful reflections and data loss will ensue. Similarly, solder joints and crimp connections must be made correctly at all points, and any variables must be closely watched.

Coax installation is easy. If you dispense this cable from the original carton or reel, it spools out nicely without tangling or twisting. Like all wire, it should not be pulled off a reel that is standing vertically on the floor. This will generate twists in the cable that will become more concentrated as you progress. Instead, put the reel on a dowel or rod, suspended between two supports such as stepladders, so that the reel rotates as the cable is pulled from it. Done in this manner, it is easy to fish coax through hollow spaces because it has just the right stiffness to be workable.

There are two common types of coax, the second of these preferable for most jobs:

- RG 59—Used for short distances only, such as the jumper cable from a video player to a TV.

- RG 6—Exhibits less loss at high frequencies and is good for longer distances.

Coax can be cut using any appropriate tool such as large diagonal cutters or tin snips. A hacksaw works, but metal filings are generated, and they must be avoided in any area where printed circuit boards are going to be present. If the end of the coax becomes elliptical or out of shape, make it round with pliers, and remove any burrs or ragged edges. Do not perform this procedure or move the coax if the other end is connected to any equipment that is powered up. A short will destroy valuable semiconductors at the output of the connected equipment.

Using a coax cable stripper, prepare the end. This tool simultaneously makes both of the required cuts to the correct depth and spacing, without nicking the center conductor. Spin the cutter three or four turns and remove the severed pieces of jacket, insulating material, and shielding. Fold the remaining shielding back over the jacket. The way you could go wrong in this procedure would be for the shielding to end up in contact with the inner conductor, which would short out the signal. Make sure that there are no loose or broken pieces that could do this.

With a twisting motion, insert the prepared end into the connector. Looking into the open end of the connector, you will see that the insulating cylinder is driven in all the way, and that the cut-off end of the inner conductor is flushed with the end of the connector, centered and not bent or buckled and not in contact with the metal part of the connector.

Lay the connector in the crimping tool and squeeze the handle. The old die hexagonal crimper simply squeezed the connector onto the cable, and the connector tended to come loose later. The new type of tool compresses the connector longitudinally, producing a long-lasting connection that seals out dirt and moisture.

The cable is connected at the equipment by screwing on the threaded part of the connector, which turns freely so that the cable will not have to twist. The connector should be put on by hand, not using a wrench, so that it can be removed easily if necessary. Do not turn or put strain on the fitting that is built into the equipment, as this will damage the electronics inside.

Two cable ends with connectors can be joined by means of an inexpensive threaded coupling, known as a barrel. Connectors can also be attached as an input or output to a splitter, allowing reception by TVs in different rooms or buildings. However, this does not work for Internet access systems, where any number of separate branches may be derived from the ports of Ethernet hubs.

NEC requirements that relate to coaxial cabling are found in Article 820, Community Antenna Television and Radio Distribution Systems. Some of the mandates are identical to those for Ethernet cabling. For example, abandoned cable is to be removed. In addition, the plenum-riser-restricted use hierarchy is applicable. These requirements affect the use of other types of low-voltage wiring as well, including fire alarm, telephone, and fiber optic cable. Additionally, the point is made that fire stopping must be used where firewalls are penetrated.

In a building, coaxial cable that is on the surface of walls and ceilings must be secured and supported so that the cables will not be damaged by normal building use. Unlike wiring for power and light, exact intervals for this hardware are not given.

Grounding is the foremost protective measure against damage due to lightning. In installations where coaxial cable enters a building, a primary protector is used to ground the coax outer shield. This piece of hardware mounts on the surface of the building, inside or out. It consists of threaded male fittings onto which coax connectors can be screwed, one from the outside cable and one from the inside cable. The primary protector has a metal lug, electrically connected to the coaxial cable outer shields, to which a grounding/bonding wire can be attached. The primary protector is to be located where the coax enters the building, not near combustible material. The grounding lug provides means of bonding the coaxial cable outer shields to the building's electrical grounding system, and the grounding electrode conductor should be sized out in accordance with NEC Article 250, Grounding and Bonding, and in no case smaller than 14 AWG but not required to be larger than 6 AWG (see Fig. 10-2).

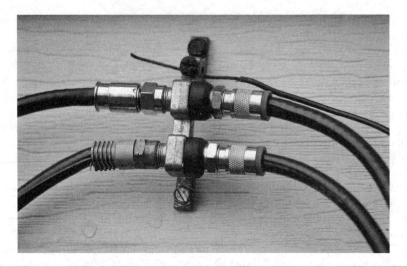

Figure 10-2   Satellite dish primary protector on an outside wall.

The bonding conductor should not be longer than 20 ft. This is a requirement for residential work, but is recommended for all locations. If the distance to a point of connection to the building's electrical grounding system is more than 20 ft, a local grounding electrode, usually a ground rod, is needed, and in that case the satellite dish grounding system is still to be bonded to the building's electrical grounding system by means of a 6-AWG minimum bonding conductor (see Fig. 10-3).

Many satellite dish installers and repair technicians fall down on their grounding procedures so that there is not adequate protection against lightning hazards. If a local ground rod is used, bonding to the building's electrical ground system is essential to prevent dangerous differences in voltage potential.

Figure 10-3   6-AWG insulated ground wire clamped to a water pipe.

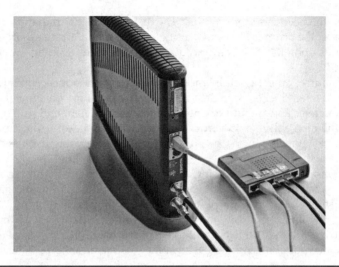

FIGURE 10-4   Internet satellite dish modem with coax input and output connected to an Ethernet hub.

After the coaxial transmission line enters the building for an Internet access satellite dish system, it attaches to the modem ("modulator-demodulator"), which processes the transmit and receive signals going in both directions. Besides modulation and demodulation, the modem performs other functions. From its power supply, it provides a dc voltage for the LNB, part of the dish assembly. It contains a network interface card and there is an Ethernet port for connection to the Ethernet hub or computer. It also has LED indicators, which greatly facilitate troubleshooting. On the latest HughesNet modem, there are five LEDs, all of which are ON when the system is operational (see Fig. 10-4).

The top LED is marked LAN, and when it is ON you know that the local area network is connected. The second and third LEDs from the top are marked Transmit and Receive, respectively, and they indicate that these functions are operating. They blink steadily when the system is idling. Data reception or transmission is indicated by a rapid, irregular blinking. The fourth LED is labeled System, and when it is on there is connection to the dish. The fifth LED is marked Power, indicating that the modem is connected to a live ac receptacle.

Dish System Inner Workings

If a discarded Internet access satellite modem becomes available, take it apart to see how it works. You should be able to see the power paths into and out of the power supply, how biases are provided for semiconductors, and find the output connections including the NIC.

An Ethernet port enables connection to the computer or, if there is more than one, to the Ethernet hub. This circuitry conforms to our discussion of Ethernet systems in Chap. 9.

Only a finite amount of energy is available for the transponder to transmit its signal to the end user's dish on Earth. The signal becomes weaker as it travels farther from the satellite because the cross section of its conical beam gets larger and, therefore, the microwave radiation becomes more rarified. The dish reflects that radiation and because of its parabolic shape, the beam is focused at a single point so that the feedhorn can pick it up. All of this depends upon the dish being accurately aimed at the point in the sky where the satellite is located.

The dish, of course, must be solidly mounted or, no matter how accurately it is initially aimed, it will not remain locked onto the satellite. If it is mounted on a metal pipe, that pipe must be set in concrete. The amount of concrete must be sufficient and its depth great enough so that the pipe will remain vertical. Heavy winds can exert a lot of pressure on the dish, and this force will try to move the pipe out of plumb, negating the installer's best efforts to aim the dish accurately.

Roof or wall mounting involves using lag screws to attach the mounting bracket to the building, and the plan must always be to have these screws go into substantial framing members, not just into the building's plywood or board sheathing. Besides being firmly mounted so that it will never move, the vertical axis of the dish must be plumb, that is, perpendicular to a level line that is tangent to Earth's surface. If this alignment is not correct, it will not be possible to aim the dish at the satellite based on settings obtained for your location.

Once it has been ensured that the dish is firmly and permanently mounted in place and that it is plumb, the installer will undertake to aim the dish at the satellite. To do this, it is necessary to know the elevation and azimuth settings that are needed. The tech help person will provide this information, or it may be obtained from a number of websites. This generally involves typing the zip code into an online form.

For purposes of this discussion, it is assumed that you are doing a repair rather than a new installation, although the procedure is the same. The dish may have been knocked down or thrown out of alignment by heavy winds, or inadvertently moved. During a reroofing operation, the dish could have been taken down and then screwed back in place, hopelessly misaligned.

Begin by setting the elevation of the dish. Loosen the bolts and move the dish until the pointer lines up with your elevation figure. If the dish was moved accidentally, the elevation may not have been changed, but it should be checked.

The next step is to set the azimuth, and for this, a compass is required. At the location of the dish, rotate the compass so that the pointer lines up with the north mark on the compass dial. Referring to your azimuth number, find it on the compass dial and point the dish in that direction. The factory setting for polarization—skew—should not be changed.

These are the preliminary settings for the dish. It is essential to begin by pointing the dish in this way. You cannot just sweep the sky, hoping to find a signal. Any attempt to do so will probably result in locking in on the wrong satellite and, of course, that will not work.

The next step is to peak out the signal. The more professional way to do this is to use a specialized meter connected at the dish using coaxial connectors. Again, it is to be emphasized that no connections should be changed while the system is powered up because if the inner conductor is grounded out, the resulting short circuit will damage the modem for an Internet access system or the receiver for a TV system.

The other less professional but viable way to peak out the system is to use the on-screen signal strength meter. Set up a laptop or small portable TV, as the case may be, where you can see it from the location of the dish. Otherwise, have a helper watch the screen inside, communicating by cell phone if possible. In this method of peaking out the system, it must be remembered that there will be a delay while the meter hooked directly to the dish provides instantaneous results. If you use the on-screen meter, you have to move the dish in small increments and wait until the meter catches up.

When peaking out a dish, strive to get the highest possible signal strength reading. In good weather there will be no difference, but in bad weather there will be noticeably less rain fade for an accurately aimed dish. An Internet access dish requires more accurate aiming than a TV dish. For either one, it is worthwhile to expend whatever time and effort are necessary to get the best possible result.

Troubleshooting Dish Systems

Now that you know the overall structure of a satellite dish system, you should be able to devise a program of troubleshooting that is appropriate to the symptoms. Satellite dish systems are easy to work on, compared to many types of electrical equipment, because they divide into discrete subsystems with readily accessible test points. For example, in an Internet access satellite dish system, the LED array on the modem divides the system in half, so that you know where to go next. Similarly, the Ethernet hub has LEDs indicating power and signal status. If either of these loses ac power, the system will be dead.

Satellite TV and Internet access systems from the various manufacturers have extensive on-screen diagnostics. For a start, they will let you know whether the TV or computer is connected to the dish and, if so, the relative strength of the signal at the dish. In addition, the system operators' websites contain much troubleshooting information that is specific to the system. If there is a weak signal or no signal at the dish, it is possible that the service provider is experiencing an outage. If you call the technical assistance line, the technicians will run a diagnostic check from their end and, very often, this will resolve the problem.

If there is a weak signal, local weather conditions can be the cause. In a TV satellite dish system, you will see pixilation, which is a way of making noise less noticeable, as a static picture is retained in the memory to help bridge gaps. Microwave transmission does not like water in any form. Heavy cloud cover, fog, rain, or snow will take down a system on a temporary basis, and the outage may be intermittent until the weather clears. In addition, an accumulation of snow or ice on the dish will cause loss of signal. Over the years, trees grow. New foliage can affect system performance. This problem will first appear in the spring when there is rapid new growth, or on a windy day when usually unobtrusive vegetation may begin to block the line of sight to the satellite. New building construction may also affect reception.

Wind can affect the aiming of the dish, which is why installers attempt to mount the dish as sturdily as possible. If the dish is mounted on the roof of an old building, where framing is weak or spongy, loss of aiming can occur.

There may come a time when you believe the LNB is faulty. It can be checked in several ways. The simplest and most reliable test is to install it into another working dish assembly of the same make and model. Another test is by means of on-screen diagnostics. Consult the user's manual for specifics. A radio-frequency power meter will measure the signal coming out of the LNB. External power is required. A less expensive instrument, an LNB in-line tester, connects between the LNB and the receiver. The receiver supplies the dc voltage for the LNB, and this current flows through the meter.

An LNB is a sensitive piece of equipment because it deals with such high frequencies and performs several operations. Moreover, it is exposed to a very wide range of ambient temperatures and there is the possibility of moisture infiltration, so it is a logical place to look if it seems like the dish has an unobstructed link to the satellite, and there is no line fault but there is no signal at the receiver.

Often there is a transmission line problem. If the dish is ground mounted on a pole, there will be an underground line. The original installer may have buried it just below the sod, so it is vulnerable to garden tillers and landscaping operations. Frost action at either end can disrupt line integrity. Roof- and wall-mounted dishes have the coax attached to the outside of the building, so ice and wind can inflict damage.

Lines can be checked with an ohmmeter. Disconnect both ends and look for a high reading, then shunt the far end and look for a low reading (not the characteristic impedance!). The ohmmeter test is not 100 percent definitive because it will not detect a subtle fault that affects high-frequency performance, but it will reveal breaks or shorts in the line, which are the usual scenarios.

The final link in the chain between a TV satellite dish and the TV is the receiver. If you suspect that the receiver is at fault, check all plug-in terminations by working them in and out gently. Then try the reset procedure. This will restore the factory defaults and additionally the receiver will download a complete new set of software, overwriting the existing data, which may have acquired a corruption.

A reset may be done by unplugging the receiver power cord, waiting 10 seconds, and then plugging it back in. Some receivers have a door, behind which are a black power button and a red reset button. You can hold the power button in for 10 seconds and then release it, or just momentarily press the reset button.

If you call tech help for a TV satellite dish system, the first thing they will ask you to do is a reset, as described previously. If you call tech help for an Internet access satellite dish system, the first thing they will ask you to do is to go to the diagnostic screen on your computer, which is done by entering the system's Internet protocol address in your browser window. The next thing they will ask you to do is to see if all five LEDs on the modem are lit. Somewhere along the line, they will ask you to restart the modem. Unplug the modem and, in this case, wait 30 seconds, and then plug it back in. It will take a while to completely reboot. When all the LEDs are back to normal, the process is complete and often this restores system operation.

The modem restart is not the same as a reset. The reset is done by connecting a Windows-based computer to the modem, and entering the MS-DOS mode. Then type TELNET 192.168.0.1 1953. Press ENTER. Then type RF and press ENTER again. The modem will restart with factory settings. However, you may have a problem getting your computer to work with the modem, and then it is a job for a factory-certified technician, so the above procedure should only be undertaken if you are aware of the potential costs.

The best course of action is to contact the tech help line and find out in advance what work they will perform and what the costs would be.

CHAPTER

Elevators

Our ancient ancestors in Africa and Asia, and later in Europe, built some very large structures, and undoubtedly they employed cranes to lift and place heavy stone materials. The cranes, powered by oxen or other animals or by humans in a turnstile, would have been framed of wooden timbers, which eventually decayed, leaving no trace. It seems obvious that these cranes would have been used to lift workers to their assigned locations, some to great heights. They must have been carried up and down, to and from work, in baskets or cages. These were the cars, and the workers were elevator passengers.

Our elevators are much faster, more numerous, and safer. When you consider how many passenger-trips are made each day, the few accidents that occur, tragic though they are on an individual level, are not frequent enough so that an elevator ride could be considered at all hazardous.

A good part of any modern elevator is electrical in nature, and where it has jurisdiction, the National Electrical Code sets standards for that portion of the installation from a safety perspective. When large numbers of people are being transported vertically in small enclosures with doors that open and shut automatically, it is easy to see that there could be the twin hazards of shock and fire. Closely adhering to the Code mandates for elevator electrical design and installation will greatly minimize the hazards. In any repair operation, especially if alterations are to be made, it is essential to conform to National Electrical Code (NEC) requirements as well as those of the Safety Code for Elevators and Escalators, ASME A17.1.

NEC Article 620 is titled "Elevators, Dumbwaiters, Escalators, Moving Walks, Platform Lifts and Stairway Chairlifts." Our focus, in this chapter, will be on elevators and their subsystems.

To begin, there are a dozen definitions of terms as they apply to Article 620. Some of these are as follows:

- Control rooms and control space are differentiated. The distinction is that the control room is intended for human entry, while a control space may be accessed without entering it. Both control room and control space contain the elevator motor controller and electrical or mechanical equipment used directly in connection with the elevator, but not the electric driving machine or the hydraulic machine.

- Control system is defined as the overall system governing the starting, stopping, direction of motion, acceleration, speed, and retardation of the moving member.

- Motor controller is made up of the operative units of the control system comprised of the starter device and power conversion equipment used to drive an electric motor, or the pumping unit used to power hydraulic control equipment.

- Machine room is defined, for purposes of Article 620, as an enclosed machinery space outside the hoistway, intended for full bodily entry, which contains the electrical driving machine or the hydraulic machine.

- Operating devices include car switch, pushbuttons, key or toggle switches, or other devices used to activate the operation controller.

- Signal equipment is made up of audible and visual equipment such as chimes, gongs, lights, and displays that convey information to the user.

If these definitions are kept in mind, our discussion of NEC elevator coverage will be easier to follow. One of the crucial things to keep in mind about an elevator repair or upgrade job is that the proper wiring methods must be used, whether in the machine room, the hoistway, or in the car. Electrical metallic tubing (EMT) in conjunction with metal-clad cable (MC) is the wiring method of choice, although some hazardous locations will require rigid metallic conduit (RMC) to ensure that flammable gases or liquids will not pose a threat to life and property.

Following the definitions, Part I, General, provides some specific requirements applicable to all elevator installations.

A voltage limitation is set at 300 volts between conductors for the supply voltage, except as otherwise permitted for power circuits, lighting circuits, and heating and air-conditioning circuits. Upper limits in some of these instances are increased to 600 volts and even higher for internal wiring of power conversion equipment if the equipment is listed for these higher voltages.

Elevator Wiring Protection

It is stated that all live parts within specified areas of the elevator installation, such as in hoistways, be enclosed to protect against accidental contact. Working clearances are to be provided about controllers, disconnecting means, and other electrical equipment as specified in NEC Article 110, Requirements for Electrical Installations, with exceptions for locations where conditions of maintenance and supervision ensure that only qualified persons examine, adjust, service, and maintain the equipment.

Part II, Conductors, discusses traveling cables. It states that within them, conductors that are 20 AWG copper or larger may be connected in parallel to obtain necessary ampacity. The advantage of this arrangement is that there is more flexibility, which is the whole point of traveling cable. This is an exception to the Code ruling on paralleling conductors, which states that, in general, this type of connection is permitted only for conductors 1/0 AWG or larger.

Section 620.13, Feeder and Branch Circuit Conductors, contains some provisions that profoundly affect the whole business of wiring elevator installations. Conductors supplying a single elevator motor are to have an ampacity not less than the percentage of motor nameplate current as shown in Table 430.22(E). This table, in Article 430, Motors, Motor Circuits and Controllers, gives nameplate current rating percentages for various classifications of service where the duty cycle is limited. One of the classifications includes freight and passenger elevators. For a continuous rated motor, the permitted nameplate

FIGURE 11-1 Elevator motor with cable drum.

current rating is 140 percent. In other words, the supply conductors for an elevator may be smaller than for motors that are presumed to run continuously, as is the general situation (see Fig. 11-1).

Furthermore, an additional reduction is permitted for feeders supplying multiple elevators. The demand factors range from .95 for two elevators on a single feeder to .72 for 10 or more elevators on a single feeder. This reduction is permitted because, as the number of elevators increases, the probability of all of them running simultaneously decreases.

Part III, Wiring, covers the wiring methods that are permitted in various parts of an elevator installation. In general, all elevator locations, excluding the traveling cable, must be installed in RMC, intermediate metal conduit (IMC), rigid nonmetallic conduit (RNC), wireways, Type MC, mineral-insulated cable (MI), or armored cable (AC), unless otherwise permitted in hoistways, as follows:

- Cables permitted for Class 2 power-limited wiring.
- Flexible cords and cables that are components of listed equipment and used in circuits operating at 30 volts ac (42 volts dc) or less in lengths not over 6 ft.

- The following wiring methods, in lengths not exceeding 6 ft:

    (1) Flexible metal conduit.

    (2) Liquid-tight flexible metal conduit.

    (3) Liquid-tight flexible nonmetallic conduit.

    (4) Flexible cords and cables that are part of listed equipment, a driving machine, or a driving machine brake.

In cars, in addition to the wiring methods that are permitted throughout elevator installations listed above, flexible conduits are allowed in less than 6-ft lengths. With conditions, flexible cords and cables are also permitted on elevator cars. Within machine rooms and control rooms, the wiring methods permitted throughout elevator installations are available alternatives. Additionally, the flexible conduits and flexible cords and cables are allowed with conditions.

The point to be remembered in all of this is that EMT and MC are the commonly used wiring methods permitted throughout an elevator installation. Under no circumstances is Type NM (trade name Romex) appropriate in elevator work. Flexible conduits, cords, and cables are allowed, mostly in lengths not over 6 ft. They are used where flexibility is needed. Traditionally, car doors have been opened and closed by means of a fractional horsepower induction motor mounted atop the car, operating the door opening/closing mechanism by means of a V-belt drive. A motor deployed in this way will usually be powered by means of a short, flexible cord, and there is no reason why it should have to be over 6 ft long.

The wiring method in each sector of an elevator installation is to be selected based on differing priorities. For example, in a hoistway, if the car were to contact wiring attached to the inner wall, the result could range from an inconvenient outage to a catastrophic event. Therefore, wiring in a hoistway should be in metal raceway and firmly anchored. Remember that repeated exposure to vibration from the passing car can loosen the securing hardware. Any maintenance or repair work in this area should include a close inspection of the existing installation in order to spot any potential problems.

Specialized Requirements

Section 620.23, Branch Circuits for Machine Room or Control Room/Machinery Space or Control Space Lighting and Receptacle(s), makes several points that, if observed, will definitely improve reliability of the installation. One requirement that should be observed in all elevator work is that a separate branch circuit is to supply the machine room lighting. The reason for this rule is that if there is an overload or short circuit in the elevator motor or controller wiring, reliable lighting that will not be simultaneously compromised is needed in the machine room for repair work to proceed. Along these lines, it would be bad practice in the course of repair or alterations to power any nonlighting equipment from the dedicated lighting circuit. Moreover, the required lighting is not to be GFCI protected, lest nuisance tripping should put the machine room in darkness at a critical time as during an elevator outage.

For purposes of maintenance and repair, at least one 125-volt, single-phase, 15- or 20-ampere duplex receptacle is to be provided in the machine room. In contrast to the lighting, these receptacles are to be GFCI protected. Similar lighting and power requirements apply to the hoistway pit, which is the area in the hoistway below the lowest limit of the car's travel.

Section 620.37, Wiring in Hoistways, Machine Rooms, Machinery Spaces, and Control Spaces, contains an important mandate that is frequently violated. The rule is that only wiring, raceways, and cables used directly in connection with the elevator, including wiring for signals, for communication with the car, for lighting, heating, air-conditioning, and ventilating the elevator car, for fire detecting systems, for pit sump pumps, and for heating, lighting, and ventilating the hoistway are permitted inside the hoistway, machine rooms, control rooms, machinery spaces, and control spaces. This requirement is often violated, leading to a proliferation of foreign wiring that makes circuit tracing more difficult during repair operations. Under no circumstances should the elevator hoistway be used as a chase to contain power and lighting or communications (including fire alarm) wiring that is not pertinent to the elevator installation. Such practice interferes with elevator maintenance and repair and additionally, if work on that wiring should become necessary, the electricians would have to enter the hoistway.

Bonding of elevator rails to a lightning protection system down conductor is permitted, but these elevator rails are not to be used for the actual down conductor. This is in keeping with the general principle that all large metal structures within a building are to be bonded to the electrical grounding system.

Part V, Traveling Cables, details requirements concerning their installation. A traveling cable is necessary to power electrical equipment aboard the car and counterweight (as is sometimes the case) and to provide communication capability. The overriding requirement for a traveling cable is flexibility. To this end, typically it contains many small wires so it can, in the lifetime of an installation, flex a great many times without damage to the copper conductors and insulation.

Traveling cables are supported by one of the following:

- Their steel supporting members.
- Looping the cables around supports for unsupported lengths less than 100 ft.
- Suspending from the supports by means that automatically tighten around the cable when tension is increased for unsupported lengths over 100 ft.

Needless to say, traveling cables between fixed suspension points are not required to be in raceway.

Part VI, Disconnecting Means and Control, provides that a single means for disconnecting all ungrounded main power supply conductors for each unit be designed and installed so that no pole can be operated independently. This disconnecting means for the main power supply is not to disconnect the branch circuits for lighting and power in the machine room and hoistway pit, or the branch circuits for lighting, receptacle, ventilation, and heating and air-conditioning in the car. No provision is to be made to automatically close the disconnecting means. Power is to be restored only manually.

Part VII, Overcurrent Protection, makes the point that the duty on elevator motors is to be rated as intermittent.

Part VIII, Machine Rooms, Control Rooms, Machinery Spaces, and Control Spaces, specifies that elevator driving machines, motor-generator sets, motor controllers, and disconnecting means are to be installed in a room or space set aside for the purpose and secured against unauthorized access. Because this rule requires that an elevator machine room be kept locked, it is important that those responsible for elevator maintenance and repair have immediate access to the key. Since an elevator outage may mean that passengers are stranded in a car between floors, it is essential that the workers have immediate access to the machine room.

Part IX, Grounding, requires, as one would expect, that metal raceways, Type MC cable, Type MI cable, or Type AC cable attached to elevator cars to be bonded to metal parts of the car that are bonded to the equipment grounding conductor. For electric elevators, the frames of all motors, elevator machines, controllers, and the metal enclosures for all electrical equipment in or on the car or in the hoistway are to be bonded.

Further, each 125-volt, single-phase, 15- and 20-ampere receptacle installed in pits, in hoistways, on elevator car tops, and in machine rooms is to be a GFCI. A single receptacle supplying a permanently installed sump pump is not required to be a GFCI.

The foregoing is a summary of the most important NEC requirements for elevator wiring. Any addition or expansion of an elevator electrical system should comply with Article 620 in addition to general design and installation rules that appear throughout the Code. As we have noted, there are safety issues inherent in all elevator installations. For this reason, it is essential that all workers involved in this area acquaint themselves with state, municipal, or other jurisdictional regulations that pertain to this sensitive field. The sensible course of action is to comply with all aspects of all governing documents. As in every phase of electrical work, but especially in elevator maintenance and repair, it is not in the interest of the technician to work under the table because this approach invites disaster and is accompanied by great liability, both to the individual worker and to the employer.

That being said, a line has to be drawn to delineate the areas in which maintenance workers can take actions to keep the elevators running. The alternative is to call in licensed elevator technicians. If they are located some distance away or are not able to respond immediately, a decision has to be made. Several considerations need to be factored in. When public safety is at stake, expense must never be an issue. As an example of the kind of thinking that will take place, suppose that an elevator has stopped running in a high-rise hotel and a car full of passengers is stranded between floors. Do on-site maintenance workers, including licensed electricians, have the knowledge and expertise to extricate the passengers safely?

When an elevator car stops between floors, the outer doors will remain closed. The procedure is to use a key to open the outer door on whatever floor provides best access to the car. The key should be either in the machine room or in a maintenance or administrative area. If it is in the machine room, which is required by NEC mandate to be kept locked, there is a minimum of two keys that have to be located. To move quickly on this, it is necessary to know exactly how to proceed.

The door key must be inserted into an opening near the top of the door and manipulated in a certain way to open the outer door. The procedure should be practiced in advance during downtime. Once the outer door is open, the inner door can be opened manually. Here again, depending upon the elevator make and model, this may require a special technique, and should be rehearsed in advance by all concerned.

If the car has stalled close to a landing, it will be a simple matter, once both doors have been opened, for the passengers to step out. If, however, this distance is too great, it will be necessary to provide a stepladder to get the passengers out of the car. At this point, one or more of the passengers may be experiencing intense anxiety, fearing that as they are emerging, the car will start to move and crush them. You can assure them that this is a physical impossibility. There are redundant safeguards that are designed to prevent this from happening. You can further assure them that the car cannot fall, they will not run out of oxygen, and help is on the way. The correct procedures are to be reassuring but in no way coercive and, if necessary, call in emergency workers to assist.

As you can see, passenger extrication requires knowledge, expertise, and professionalism. One elevator manufacturer advises on its website for maintenance workers not to do a passenger extrication, but to call a licensed elevator technician. I would say that the issues are not always that simple. If there will be a delay, medical or psychological issues could emerge. In some localities, firefighters are adept at elevator passenger extrication, and they possess a selection of outer car door keys. Procedures should be in place and rehearsed so that in an event of this sort the response will be appropriate.

The Next Step

Once the passengers are extricated, if necessary, the task of getting the elevator going again becomes the priority. Among the on-site workers, electricians are usually asked to make an initial assessment. A policy should be in place regarding how to proceed before calling in licensed elevator technicians. As a recent tragedy in midtown Manhattan demonstrated, there is danger at all levels of competence, so this has to be considered carefully. For the present discussion, we shall explore some of the intricacies of elevator technology from an electrical and mechanical perspective. To begin, we will look at the overall structure of a typical single elevator installation.

> A submersible pump of 3 hp or higher should have soldered pump cable splices. Crimp the connectors in the usual way, then apply resin flux, and run solder into the crimps.

A good way to start is to look at the power flow. The electrical supply is generally three-phase. From the entrance panel, a feeder will run to the machine room, terminating at a disconnect. This NEC-required device may include circuit breakers or knife switches with cartridge fuses, and it is often located on the wall near the door. It is usually labeled, but if not, it will be self-evident with large metal raceways going in and out. As previously noted, power for lighting and receptacles in the machine room will be on separate branch circuits that do not go through this disconnect enclosure.

From the disconnect, conductors will go to the variable frequency drive (VFD), and from there to the motor. The elevator controller, in a large enclosure, is made up of a CPU, with inputs from sensors, control switches, and the building's fire alarm control panel, and outputs to the VFD, car door operator, and whatever actuators and indicator lights exist. The control panel will have a front cover that can be removed, providing access to electronic components and the alphanumeric display. This readout indicates that the system is operating normally or, if not, an error will be displayed. The error code will be somewhat laconic, saying something like E 10 Door Interlock Malfunction. For a full explanation of all error codes, consult the operator's manual, either a print version supplied by the vendor when the elevator was originally installed or an online version downloaded from the manufacturer's website.

By far the most common cause of elevator malfunction is that a door did not close completely. Obviously, an elevator cannot permit itself to run if one of the doors is free to open because a person could be caught and crushed in the gap. This is why there is an inner door that travels with the car and an outer door at each floor, the latter serving to prevent an individual from falling to the bottom of the shaft. A door sensor reports via a simple two-wire circuit when the door is tightly closed. The car will not move if the door is even slightly loose, if the switch mount has become loose or otherwise faulty, if there is a line fault going to the controller, or if there is a problem in the controller itself. It is highly unlikely that there will be a malfunction that will permit the elevator to move with an open door (unless the circuit is purposely disabled for maintenance purposes). The single most common cause of elevator failure is that the door sensor is overzealous in performing its protective function. This is similar to a false alarm in a fire alarm system. The easiest way to restore service is to go to each floor and check the outer door. If the inner door has jarred loose with no one in the car, it may be necessary to open the outer door with the emergency key and close the inner door manually. This will restore service without restarting the control panel. Depending upon the operating parameters of the specific system, the elevator may immediately start up and go to the next floor on the agenda. However, the underlying cause (loose door hardware or sensor problem) should be ascertained so that the same situation will not recur. Under no circumstances should the door interlock be disabled.

If the door has a bottom track, it can collect dirt and debris and this may cause elevator shutdown. Similarly, the light source and receptor that are intended to detect any object, that is, a person, in the door opening may falsely report blockage when it is actually due to dirt on the optical surface, thereby preventing the automatic door from closing. This can be fixed by a simple cleaning.

When door and door sensor malfunctions have been ruled out, go into the machine room. It is understood that a fault in the power supply will put the elevator out of service. Keep in mind that the lighting is on a separate dedicated circuit, so you cannot take that as an indication that there is elevator power coming into the machine room. There must be voltage at the input and output of the disconnect. This is easy to verify. Once it is known that there is power to the controller but the elevator will not respond to call buttons at the floors or inside the car, the question becomes whether to reset the controller. There is a difference of opinion on this matter. Inside the controller enclosure, mounted on a printed circuit board close to the alphanumeric display is a very small red button labeled in very small letters RESET. Sometimes the reset button is hard to find and sometimes it is a color other than red. The temptation is to hit the reset button because nine times out of ten, once door faults have been eliminated, this will restore service. However, some technicians contend that the underlying problem should be found and repaired prior to doing a reset.

The controller is a CPU with an operating system and many capabilities in some ways similar to a desktop computer. When you hit the reset button, the controller reboots, and this takes about a minute. Resetting the control panel may also be done by shutting off the disconnect, and then repowering. There are those who say in stubborn cases it is helpful to leave the system powered down for a longer period of time. The rationale for not resetting the controller has these counter arguments:

- It usually works.
- The primary diagnostic tool is the controller, which needs to be rebooted before you can do anything.
- Since time is of the essence when it comes to getting an elevator going, we need to get it back in service quickly.

For these reasons, most technicians feel that aside from a brief visual inspection for blown fuses, loose wires, or bad capacitors, a restart is appropriate. If resetting the controller does not restore the elevator to normal operation, the alphanumeric readout will display an error code. Refer to the user's manual for a description of the malfunction and recommended course of action.

Care must be taken when working inside the control panel so that the elevator safety features are not compromised. In a 2004 Occupational Safety and Health Administration (OSHA) bulletin titled "Hazards of Improper Elevator Controller Wiring," a fatal accident is discussed. An employee entered an elevator that continued to move with the door open, and was crushed between the elevator car and the hoistway. The cause of this accident was traced to an improper repair performed one day earlier. Two wires were switched and reattached to the wrong terminals, which were adjacent. This reversal inadvertently caused the hoistway door interlock to be disabled, allowing the car to move with doors open.

From this tragic event, a lesson may be learned. It is that when it comes to elevators we must never work outside our area of competence. Nobody knows everything, and whether you are a licensed elevator technician, licensed electrician, apprentice, or maintenance worker, there is the potential for making a mistake that will someday, perhaps far in the future, threaten worker or public safety. If in doubt, call in expert help, and in every case know your limitations and do not work beyond them.

The elevator controller is capable of placing the elevator in modes of service that differ from normal operation. One of these is inspection mode, and it is used by elevator technicians when initially setting up a new installation or doing maintenance or repair. Inspection mode is entered by activating a small toggle or key switch inside the control panel. The elevator will stop, if it is moving, and it can only be moved by car top or control panel buttons marked UP and DOWN. The car moves at reduced speed with doors open, allowing technicians to perform tasks that would otherwise not be possible. When running the elevator in inspection mode, access to the area should be restricted so that inexperienced workers and visitors will not be endangered.

Fire Service Mode

Another non-normal mode of elevator operation is Fire Service Mode. It is actually made up of two separate modes: Phase One and Phase Two. The basic premise is that in the event of fire, members of the public should not use an elevator to evacuate the upper floors of a building. This is because at any time, due to fire damage, the car could stall during its descent, trapping the passengers in a burning building. Instead, occupants should use the stairway, which is enclosed by firewalls and often pressurized to prevent smoke from entering.

To this end, when a smoke detector on any floor near an elevator is activated or it is so instructed by the main fire alarm system, the elevator will enter Phase One of Fire Service mode. Immediately, all elevator calls from all floors will be cancelled, and the car will go to the designated recall floor, usually the ground floor, if that is not the location of the fire; otherwise, it will go to some other predetermined floor. It will remain there with the door open and not move as long as the elevator is in Fire Service Phase One. The priority in Phase One is to get car passengers to a safe location quickly before the power fails.

The purpose of Fire Service Phase Two is to assist firefighters in putting out the fire and rescuing any other occupants of the building. Phase Two mode is entered by activating a key switch grouped with the control buttons inside the car. This switch is colored red and

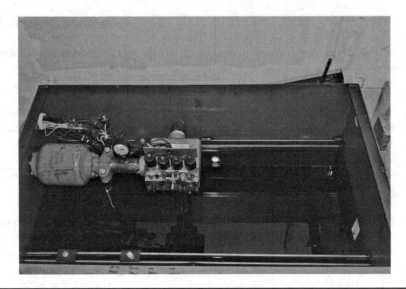

FIGURE 11-2 Oil reservoir for hydraulic elevator.

has a graphic of a firefighter's hat. After activating the key switch and entering Phase Two, the firefighters can cause the car to go to any floor by pushing the required button and simultaneously pushing and holding the DOOR CLOSE button. When the car arrives at the designated floor, the door remains closed unless the firefighter pushes and holds the DOOR OPEN button. To cause the car to remain at that floor with the door open, the firefighter turns the key switch to the HOLD position.

After the fire emergency has ended or it has been verified as a false alarm, the elevator can be placed back in the normal mode by turning the key switch at the ground floor or designated fire recall floor first to BYPASS, and then to OFF. (This key switch is also red, with a firefighter's hat.) If the switch is lighted, it means that the elevator is still in the Fire Service mode and that it has to be reset before the elevator will run in the NORMAL mode.

There are two major types of elevators—traction and hydraulic. Traction elevators have a spool at the bottom adjacent to the motor inside the machine room. It winds in the cable to make the car go up and lets out the cable to allow the car to go down. The cable goes up to an idler sheave above the top floor, and then down to where it attaches to the car.

A more advanced type is the hydraulic elevator, which however is limited to use in buildings not over four stories high. As part of a renovation, many old traction elevators are converted to hydraulic elevators, which involves drilling under the hoistway pit to permit insertion of the cylinder. The ram is attached to the bottom of the car, extending out of the cylinder to push the car up and retracting to allow it to descend. The electric motor drives a hydraulic pump and the pressurized oil forces the ram upward, while an electrically actuated valve relieves the pressure, allowing the car to move down at a regulated speed.

There is a large oil reservoir (see Fig. 11-2) in the machine room and on a busy day with high ambient temperature, the oil may overheat, causing the elevator to shut down. The remedy is to add an oil cooler or improve ventilation or air-conditioning in the machine room.

Fire Alarm Systems

Any electrician who does commercial work will quickly encounter a fire alarm system, and this can be a humbling experience due to the characteristic complexity and sensitivity of this equipment. Most buildings other than residential occupancies have fire alarm systems, installed either when the structure was originally built or retrofitted. Older fire alarm installations are known as legacy systems, and while they may provide good fire protection, they may be harder to work on and tend to give false alarms, which are annoying in the extreme, and dangerous in the sense that occupants come to disregard the real thing.

Modern fire alarm systems are integrated with the sprinklers and elevators, in addition to ventilation, automatic fire doors, gas supply, telephone, and other building services so the complete system, especially in a large building that is open to the public, is complex and multi-faceted. To the electrician or maintenance worker, some aspects may seem bewildering. The system may be difficult to maintain, especially in an occupied building where a deranged system may be giving frequent, loud, disruptive false alarms.

> Sometimes when a large piece of machinery has a noisy bearing, it is difficult to determine which one is the culprit. You can use a long screwdriver as a stethoscope. With your ear against the handle, touch the tip to the outer housing of each bearing to pinpoint the problem.

We are not talking about individual smoke detectors used in homes and apartments, typically powered by 9-volt dry cells and hardwired to the ac electrical supply, even if these units are wired together to sound an alarm in concert. A fire alarm system found in commercial buildings is a complete integrated system of great functionality. It is characterized above all by the fact that all parts are connected to a central control panel, and supervised at all times. The supervisory function is strictly electronic, and extends to all parts of the system at all times. We will look at how this is accomplished by means of an ingenious wiring arrangement that is common to all makes of fire alarm systems worldwide.

Fire alarm systems at both design and installation level are covered by multiple documents, which are generally enacted into law by local jurisdictions. Those electricians interested in expanding into this line of work will want to become familiar with the following:

- The Life Safety Code (NFPA 101) lists the types of occupancies where fire alarm systems are required. A blanket requirement for fire alarm systems in all buildings may be on the horizon. This would be a great lifesaver, although the expense would be substantial for homeowners.

- National Fire Alarm Code specifies fire alarm system design, including minimum performance requirements, operating details, maintenance and testing procedures, and individual installation requirements such as location and minimum distances between smoke detectors, horns, pull stations, and other devices.

- National Electrical Code (NFPA 70) is applicable except where modified. Specific requirements appear in Article 760, Fire Alarm Systems. This article includes information on zone wiring, sprinkler water flow and supervisory details, smoke door and fire door release, fan and gas shutdown, elevator fire service mode, and so on. Other National Electrical Code (NEC) articles are applicable, notably Article 725, Class 1, Class 2 and Class 3 Remote Control, Signaling and Power-Limited Circuits, when the fire alarm system is designed to operate at reduced voltage and power levels, which is usually the case.

- Underwriters Laboratories and similar inspection organizations list fire alarm system components, such as smoke detectors, control panels, pull stations, horns, batteries, end-of-line resistors, and the like.

Electricians frequently deal with wood framing members in both new construction and renovation. Wood framing members under load can be in compression or tension. Compression is when the piece is under load and the weight of the load is compressing it, as for example an upright post supporting a carrying timber. Tension is just the opposite, as when the piece is being stretched. A horizontal joist, supported at both ends and carrying a floor load, is in compression along the top edge and tension along the bottom edge, these stresses being at the maximum midway between supports. If you must drill holes to run cable, locate the holes midway between top edge and bottom edge where the piece is in neither compression nor tension, and offset as much as possible from midway between supports to avoid compromising the framing. It is highly undesirable to notch such a beam at the edge.

Fire alarm systems from various makers such as Simplex, Edwards, and Honeywell are substantially similar in terms of inner workings and operating details, although the parts are not interchangeable and the programming methods vary. These and other manufacturers provide excellent installation and user's documentation in the form of print and online manuals. They can be copied and kept in the shop for ready reference. There are online courses that are designed to prepare technicians for state and local certification.

Licensing requirements for fire alarm design and installation vary widely among different jurisdictions. Some states have separate classes or levels of professional fire alarm technician certifications, while others do not regulate this area at all, although presumably those engaged in the trade would be licensed electricians.

Fire Alarm System Fundamentals

Besides the control panel, there are two main categories of equipment that make up the fire alarm system: initiating devices and indicating appliances. As the name indicates, the initiating device reports to the control panel when it detects products of combustion. We immediately think of smoke detector heads, as seen spaced at uniform intervals in public places. (The stand-alone battery and ac-powered units that are used in residential applications emit an audible tone and are not connected to a supervisory control panel. They are properly called "smoke alarms.") Smoke detectors (see Fig. 12-1) that are part of an integrated fire alarm system are either photoelectric, operating by means of optical detection, or ionization, capable of detecting the physical properties of smoke that enters the detection chamber. For shops in maintenance areas where normal activities may generate smoke or dust, heat detectors are used so as not to set off false alarms.

Other types of initiating devices include pull stations, prominent red-colored enclosures mounted at strategic public locations and intended to be manually operated in the event of fire. Some of these have small glass windows that must be broken prior to activation to discourage the casual prankster.

Another initiating device is the optical beam detector. It is composed of a transmitter that emits an invisible infrared beam directed through the area to be protected, generally high up near the ceiling to avoid false alarms. There is a built-in time delay so that an alarm will not be set off by a paper airplane or similar passing event. Optical beam initiating devices provide excellent protection and are an economical means of covering very large areas where the cost of many heads would be substantial. The two types of optical beam smoke detectors are projected and reflected. In a projected assembly, the transmitter and receiver are located at opposite ends of the area that is to be protected. In a reflected assembly, transmitter and receiver are built into a single unit, known as a "transceiver." A reflector is mounted at the far end of the protected area. Both types can provide line-of-sight coverage of spans of over 300 ft, so in a large square room, several parallel assemblies would provide good coverage. The projected

Figure 12-1 Smoke detector on a finished ceiling.

beam smoke detector is less susceptible to stray reflections because the reflections do not reach the receiver. The reflected beam smoke detector does not have this advantage, but it requires less wiring. Optical beam detectors, like discrete heads, do not work effectively outdoors, where any breeze will disperse the smoke, but beam detectors are useful in hazardous areas.

To fish cable between two holes in wall material within the same framing pocket, drop an iron bolt tied to the end of a piece of string from the upper hole. Using a magnet attached to a dowel, capture the bolt inside the wall and draw it out the lower hole. This creates a pull rope, which can be used to pull the cable.

The other major category of equipment that is connected to the fire alarm control panel is known as the indicating appliance. Classic examples are large, very loud horns that used to be seen in public places. They are being replaced by electronic sounders that are less strident but equally effective. These units also contain a very visible strobe light to protect those who are hearing-impaired.

The initiating devises are daisy-chained, that is, wired in a parallel string for each zone. The building is divided into zones based on locations, and given names within the control panel, such as Fourth Floor, North Side. In a legacy system, the control panel will report only the zone of the alarm, not the exact head. The newer addressable-head systems pinpoint on the alphanumeric display the individual head or other initiating device that has activated (see Fig. 12-2).

Initiating devices are spaced at prescribed intervals along a two-wire line in the simplest system. Each smoke detector base has terminals for two wires coming in and two going out to supply the next unit. Bases are attached to octagon boxes or other enclosures mounted prior to installation of wall or ceiling finish.

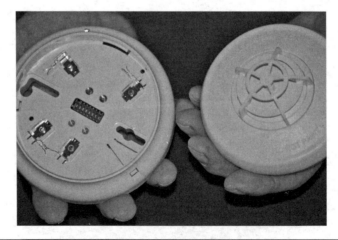

FIGURE 12-2 Addressable head smoke detector with base.

The normal state for a smoke detector is nonconducting. When smoke or dust enters the detection chamber, the head becomes conducting, the control panel sees this altered state, and the system goes into alarm activating the indicating appliances.

The wiring for initiating devices is generally 16 AWG copper, available as two conductors in a red jacket and known as fire alarm cable. This size represents a reduction from the general NEC minimum conductor size of 14 AWG. The fire alarm is generally run in electrical metallic tubing (EMT) for enhanced protection and ease of pulling out and reinstalling new wires if that becomes necessary. These conductors carry a low dc voltage, one side positive and the other negative, both isolated from ground. The metal raceway is grounded by virtue of the connectors attached to the control panel, which is in turn grounded by the equipment-grounding conductor that is part of the ac supply wiring. This grounding extends throughout the zone because the runs of metal raceway are connected to the metal enclosures for each smoke detector base.

The low-voltage dc carried by the line serves two purposes. It provides biases for the solid-state devices within the heads, and it makes possible the supervisory mechanism whereby the control panel knows what is going on throughout each zone.

The control panel is usually in one of three states—normal, alarm, and trouble. This is a very ingenious system where it is virtually impossible for the system to fail to go into alarm in the event of a fire, without first going into the trouble state, alerting workers to repair the fire alarm system while initiating human fire patrols of all areas not covered. If the system is in the normal state, the alphanumeric display will so indicate, there will be no alarm, and the fire alarm system will maintain its vigil. If the system goes into alarm, indicating appliances will sound loudly throughout the building. (In certain facilities, such as hospitals and nursing homes where a full-scale evacuation is not recommended and where a loud alarm could stress vulnerable patients or clients, softer chimes may be used so that staff may take whatever steps are needed to protect the occupants.)

In case of alarm, staff should follow procedures that have been established in advance. One or more workers will immediately go to the control panel and consult the alphanumeric display to find the location of the affected initiating device. Simultaneously, this worker should press the touchpad button labeled ACK, which stands for Acknowledge. This will cause the control panel to enter into the history the date and time when the alarm was acknowledged, which could become important later for diagnostic, insurance, or legal purposes.

If a motorized valve in a steam line has the right voltages but the motor will not turn, it is possible that the packing nut is too tight. Loosen it one-quarter turn and see if the valve becomes functional. Another possibility is that the gear train has become corroded and is sticking. Open the box and clean and oil all moving parts. Use an ohmmeter to see if the switchgear is working. These units are very expensive and worth spending some time on.

After ACK is pressed, the alarm will continue to sound. When facility workers reach the affected zone, they need to find the source of the alarm. (If it is an addressable head system, they will already have this information.) For a legacy system, they will need to travel the length of the zone. Unaffected heads will have a blinking LED indicating, as always, that they are not the source of the alarm. The affected head will have an LED that is not blinking, but on steady. If there is no fire (the alarm was caused by dust, steam, vandalism, or head malfunction), the responding worker can inform those back at the control panel so that they can silence the alarm, by pressing the ALARM SILENCE button on the touchpad. If the fire is small and localized, so that it can be put out by means of a fire extinguisher or has already been extinguished by a sprinkler head, the alarm can be silenced as well. If the fire cannot be contained, the system should remain in alarm, the indicating appliances not silenced, and procedures begun to evacuate the building. After the condition that caused the system to go into alarm has cleared, it is necessary to do a system reset. Then the alphanumeric display will read SYSTEM IS NORMAL.

Altered States

When the system goes into the trouble state, a low-level alarm will sound. Usually a buzzer, it will indicate to nearby workers that the system needs attention, but this alarm is not loud enough to disturb the occupants of the building or interfere with normal activities. The control panel monitors the status of the entire fire alarm system at all times, including indicating devices, zone wiring, internal control panel circuitry, backup battery voltage and ac supply, sprinkler water and air (for a dry system) pressure and alarm wiring, telephone lines to the fire department and monitoring agency, etc. If any of these mechanisms loses capability, the system will enter the trouble state and the alphanumeric display will describe the problem.

Depending upon the nature of the trouble, a single zone may not be operational, but the other zones will maintain their protective functions. If part of the system is going to be down, it will be necessary to initiate a human fire patrol, to contact the monitoring agency and fire department, and to contact the insurance company to assure them that backup measures have been instituted and to make sure that there is no lapse in coverage. Meanwhile, the priority should be to get the fire alarm system back into the normal state quickly.

When the trouble situation has been repaired, the system should revert to normal on its own. If it does not, it may be necessary to do a system reset by pressing the appropriate button on the console.

We mentioned that when the system is normal, the control panel is monitoring the integrity of the initiating device circuits at all times. We also mentioned that the initiating device is open when there is no product of combustion present, and that it conducts when there is smoke. The question arises: How does the control panel distinguish between an open line and a nonalarm state throughout the zone? The answer is by means of an ingenious arrangement whereby a resistor is placed across the two circuit conductors after the last initiating device. The value of this kilohm-range end-of-line resistor, as read by the control panel in its supervisory role, indicates that there is no open and no alarm. The end-of-line resistor goes across the output terminals in the base of the final initiating device. (In Canada, the end-of-line resistor is required to be in a separate enclosure. In Europe, an end-of-line capacitor is used. A manufacturer in New Zealand has come out with a control panel that can be programmed to look for either a resistor or a capacitor.)

Some technicians have put the end-of-line resistor "in the can"; that is, across the initiating device zone terminals inside the control panel. The system will appear to function in this way, but the line will not be protected. The system will not go into the trouble state in response to any open that develops beyond the resistor.

Similarly, some technicians install the end-of-line resistor inside the control panel to keep the system out of the trouble state while they are working on a dysfunctional zone. However, it is easier to disable the zone by scrolling through the trouble reports in the alphanumeric screen until the zone of interest appears, then press DISABLE, and then activate RETURN. This procedure will initiate a new trouble event, so it will be necessary to contact the monitoring agency.

The control panel also monitors the indicating appliance line, and it has an end-of-line resistor. The value in ohms is different from the initiating device end-of-line resistor, so the two should not be interchanged. (Fire alarm system manufacturers also employ differing resistances, so it is essential to use the correct resistor.)

We have only scratched the surface of the many features of a modern fire alarm system. It is worthwhile reading through the operating and installation manuals of various manufacturers. The documentation is available at no cost online, and it contains detailed information on installing the hardware, system programming, setting parameters and preferences, and so on. Programming is accomplished by means of keystroke combinations on the touchpad.

One interesting aspect of the fire alarm system is known as the "city tie." This is a telephone connection to the fire station, monitoring agency, emergency dispatch, or owner of the building—whatever is set up in the control panel programming. To make it work, a modem has to be purchased and connected to the control panel. Whenever the system goes into alarm or trouble, the control panel automatically dials the predetermined numbers, and there will be some response ranging from a telephone call to dispatch of firefighting trucks depending upon local policy.

There are two dedicated telephone lines and at predetermined intervals, the control panel places test calls to determine that the lines are operational. If one of the telephone lines is down, the system goes into the trouble state and the alphanumeric readout displays the information so that corrective action may be taken.

Besides the ac utility and backup supply, there is an additional battery backup. This is typically made up of four 6-volt batteries wired in series to make a 24-volt system. The batteries resemble emergency light batteries, but they must be listed for fire alarm system applications. The batteries are often located inside the control panel at the bottom. Of course, there is a battery charger. When ac power is lost, the batteries take over. If the battery voltage drops below a predetermined level, the system goes into the trouble state and the alphanumeric readout displays the appropriate message.

Throughout the building, spring-loaded fire doors are held open by electromagnets. If the system goes into alarm, the dc voltage at the magnets is interrupted, and the fire doors slam shut. Later, they can be reopened manually.

The fire alarm system interacts with the sprinkler system in a variety of ways, the result being that every sprinkler head is in effect an initiating device. To understand the full relationship, it is necessary to examine the inner workings of the sprinkler system. There are two types of sprinkler systems—wet and dry. The wet system consists simply of sprinkler heads connected to a high-pressure, high-capacity water supply. When a sprinkler head ruptures due to heat from flames below, a large amount of water at high pressure puts out the fire or, if the fire spreads, additional heads release water.

Sprinkler Varieties

The wet system may only be used where there is no chance of freezing. The dry system can be used in unheated areas. The dry system also has the advantage of providing advance notice of a potential leak by means of a drop in air pressure, while the wet system leaks water without warning. The dry system has a short delay before the water flows, so this may be a disadvantage if the fire is rapidly accelerating. Both systems require an abundant water supply, typically from a 6-in. pipe tapped from the water main or gravity fed from a large reservoir.

In a dry system, the water supply terminates at a large cast iron control valve assembly. It contains a large clapper, held shut by a combination of spring pressure and air pressure from above. Air pressure throughout the sprinkler piping and extending down into the control valve assembly to the top surface of the clapper is maintained typically at 55 psi. In combination with the spring, the clapper is held shut against the greater 150 psi typical of the water supply at the bottom surface of the clapper. If air pressure drops typically below 32 psi due to rupture of a head or pipe, the balance is tipped and the clapper flies open, allowing a tremendous flow of water into the sprinkler piping.

If the air pressure drops below typically 45 psi, due to a slow leak in the hundreds of feet of sprinkler piping, a trouble alarm will sound and an indicating panel that is part of the sprinkler system will indicate the affected zone. Additionally, an air pressure sensor that is tied into the fire alarm system will cause it to enter the trouble state, so that workers can increase the air pressure in the dry side of the sprinkler system by opening a valve that is connected to an air compressor. Alternately, a solenoid valve can be activated automatically so that air will be added. If the air pressure drops sufficiently for the clapper to open, a water flow sensor will cause the fire alarm system to go into the alarm state, with the sprinkler zone displayed on the alphanumeric readout. The sprinkler system will be flooded and will have to be drained, repairs made, the clapper reset, and the fire alarm system reset.

A number of sprinkler system conditions can be reported to the fire alarm system and can cause it to go into the trouble mode. This capability is invaluable in spotting problems before they cause a lapse in coverage or actual damage to the building. Low water on the supply side of the clapper and valve tampering will initiate trouble alarms.

A sprinkler system is a very effective firefighting tool. Water damage to a building may be one percent of what the fire damage could be for an unprotected building. For a new system, the initial investment is very high and for an old system the maintenance and repair costs are very high, but the protection for property and human life cannot be measured.

A sprinkler system integrated into the fire alarm system makes for a powerful firefighting combination, but in a system of such complexity, there is the potential for faults. They can initiate false alarms, which are aggravating in the extreme. In a public place such as a restaurant or hotel, guests may have to be evacuated or at least disturbed. In an office building or factory, the workflow will be interrupted, a costly situation, and maintenance workers will be diverted from their scheduled tasks. Worst of all, repeated false alarms may lead occupants of a building to ignore a real alarm.

The solution is a thorough, in-depth, well organized, and documented preventive maintenance program and regular upgrades. Second, when things go wrong, we need to have a knowledge of fire alarm system fundamentals and troubleshooting techniques so that repairs can be made quickly and property and lives are not placed at risk.

For a dry sprinkler system, the most basic maintenance operation is to check each control valve assembly for air pressure. The control valve assemblies are usually grouped, sometimes five or six to a location, based on the layout of the building. Each control valve assembly will have pressure gauges showing air and water pressures. If it is an old system with extensive piping throughout a large building, it will need to be checked every few hours, and air added if needed. Control valve assemblies with automatic air chargers should also be checked. They can be treacherous because a leak will not be detected and will gradually worsen until the pipe bursts, causing extensive water damage and a downed system.

A log should be posted at each control valve assembly, to be initialed with time and date of each inspection and amount of air added. These logs can be reviewed to spot lines that are losing excessive air, and bad sections of piping can be replaced. This involves shutting down the sprinkler zone and notifying the monitoring agency before the fire alarm system enters the trouble state.

As for the fire alarm system, testing of all initiating devices and notification appliances should be done according to a preset schedule. Smoke detector heads from various manufacturers are tested differently. Some manufacturers provide pressurized cans of simulated smoke, while others recommend bringing a strong permanent magnet close to the head, which will set it off. The control panel can be set to make the alarms inaudible for purposes of testing.

At the control panel, battery cables and terminations should be cleaned regularly. It is important that paperwork and old parts not be allowed to accumulate inside. The area in front and to each side of the control panel should be kept clear. There should be good lighting that illuminates the inside of the control panel when the cover is removed. A copy of the user's manual should be kept nearby, and another copy in the maintenance shop. It is a good idea to have a phone jack nearby in case it becomes necessary to speak to tech help or the monitoring agency while working at the control panel. A dedicated cabinet, near the control panel or in the maintenance shop, should contain spare parts including replacement heads and bases, zone cards, end-of-line resistors, spare terminal screws, and other items that may be needed.

Three-way switches are less confusing to install or troubleshoot if a few simple principles are kept in mind. Two 3-way switches plus any number of 4-way switches in between function as a block like one single-pole switch. Between all switches are red and black "travelers" that serve as alternate hot paths. For switched power to a fixture, a white neutral must accompany the travelers. When troubleshooting and the switches appear functional but the fixture will not power up, look for a problem in the neutral.

The most important tool is knowledge. There should be at least one electrician or maintenance worker who is very familiar with the particular system in use and fire alarm work in general. This person should schedule periodic meetings with other workers who are involved in maintaining and using the system. Most large facilities have safety committees and they should become involved in this important area.

Elaborate preparations have been made and there is a good maintenance program in place, but the time comes when repeated false alarms or trouble states are occurring. It is at this time that our troubleshooting skills are put to the test. To begin, we will look at false alarm categories.

The system goes into alarm. You go to the affected area and find there is no fire, just a head with a steady-on LED, not blinking. There are three possibilities:

- There was an event that caused dust, smoke, steam, flying fibers, or some other material to enter the detection chamber of the head, setting it off.

- There has been a gradual buildup of these materials that put the head into a conducting state, causing the fire alarm system to go into alarm.

- The head has gone bad.

Easy Remedies

The first thing to do is try to clean the head. Most heads remove from the base by rotating a one-quarter turn counterclockwise, and then pulling straight away from the base. Sometimes plastic parts that have been in contact under pressure for some time tend to stick, so it may take a little force to remove the head. It is good to examine a spare head and base in advance to see how they go together. After removing the head, blow through the opening while rotating it. Then tap its edge on a hard surface several times, again rotating it with the opening pointed down, to dislodge and remove any loose debris. Finally, blow it out one more time. Then, reattach the head with a one-quarter clockwise turn, and call back to the workers at the control panel, telling them to do a reset. They should be warned in advance that the system could again go into alarm. If this happens, after verifying that this is the nonblinking head, that is, the alarm is not coming from elsewhere (an incredible coincidence, but in this business we cannot take any chances at all), have your coworkers again silence the alarm.

Now you have a decision to make. You can try again to clean the head, or you can put in a new one. Heads are expensive, running more than ten times the cost of a residential smoke alarm, but then you cannot keep setting off alarms. That is the dilemma. (Some manufacturers accept faulty heads back in trade.) If there is a spare head on hand, you can put it in, and most of the time that will fix the problem. Later, in the shop, you can build a test stand, using an octagon junction box, base, and end-of line resistor. Using your system's parameters including source impedance, apply the correct dc voltage and measure the voltage drop with and without simulated smoke. Compare this to known good and bad heads.

Returning to the situation at hand, let us now suppose that the system went into alarm and the control panel indicated a specific zone. It is a legacy system, without addressable heads, so only the zone is indicated, not the specific indicating device. Besides the heads, other indicating devices can become conducting, causing the system to go into the alarm state. Any pull stations should be checked. Those with glass windows or plastic rods will bear signs of tampering. Some can be opened by means of a screwdriver or a key.

If the two conductors on an indicating device zone line become shorted, this will produce a false alarm. The condition is rare because the damage to the insulation of both wires would have to occur at the same place. Much more likely, both of the conductors will ground out against the inside of the metal raceway at different times and different places, or similar faults will develop where the wires enter enclosures. Any of these conditions, occurring singly, will give rise to the trouble state and the system will not go into alarm.

For either chronic false alarm conditions or trouble reports, it is necessary to check for line problems. The basic procedure is first to disable the zone. Some technicians, as we have mentioned, put an end-of-line resistor across the zone terminals inside the control panel so that the system can be made to return to normal, without a trouble signal. However, this practice is noncompliant because the zone is not connected. Either way, there is no protection and human fire patrols should be commenced and maintained until full coverage is restored.

Remove both zone wires from the control panel zone terminals. Then, disconnect the conductors from a base midway through the zone. Take continuity readings with and without the wires shunted at the other end. Also, take ohm readings between each conductor and the metal raceway. These readings should be in the very high megohm range—essentially open circuits. In this way, using the half-splitting technique, you should be able to locate the fault.

To summarize, a short between the two conductors, which is highly unlikely when they are in EMT, will give rise to the alarm state. A fault to ground of one of the conductors will force the system into the trouble state, and this is much more common. It will be indicated at the alphanumeric display.

The grounded conductor fault occurs frequently as an intermittent. Unfortunately, it cannot always be located by means of an ohmmeter. That is because the ohmmeter 3-volt test voltage is not sufficient to ionize a conductive path through the damaged insulation. For other types of electrical equipment, such as an electric motor, a high-voltage megger test is performed for a prescribed period. This is definitely not recommended for fire alarm systems because the testing process stresses the insulation and may create a new fault or lay the groundwork for an arc to ground in the future.

The best way to go is to wire fully charged 6-volt batteries in series to make up your fire alarm supervisory voltage, as measured at the initiating device zone output terminals in the control panel. Make a convenient carrier for these batteries and connect leads to them with alligator clips. Then you can apply this voltage to the zone wiring at various points and take voltage readings between the conductors and between each conductor and the metal raceway. Do not perform this test with the far ends shunted or the sparks will fly.

The method described above is the only safe way to isolate some subtle fire alarm system line faults that will initiate trouble or alarm states. A wire that is only slightly pinched at a connector or enclosure entry can cause this condition, whereas a similarly damaged conductor might never be a bother for a telephone circuit, to mention one example. Fire alarm systems are very fussy when it comes to their wiring.

When the faulted cable run has been identified, it should be replaced. If the circuit has been installed in EMT, this is a simple matter, using the old cable as a pull rope.

Other trouble conditions appear on occasion. As part of its supervisory function, the control panel monitors its own internal circuitry. A frequent trouble state, as reported in the alphanumeric display, is "Bad circuit card." The system will continue working with a bad card, but the associated zone will be down. The cards are in dedicated slots, which should be labeled. Just pull out the old card and put in a new one. Take the usual precautions against static electricity damage. Hold the card only at its edges, and do not touch any traces, terminals, or conductive surfaces. Touch the grounded control panel enclosure periodically so that any static charge in your body will be shunted to ground. If a copper grounding bracelet is available at this location, make use of it. A static charge is retained to a greater degree when the surrounding air is dry, as in a heated building in winter, and it is dissipated more readily when the air has greater water content as on a humid day in summer.

If a situation arises where the fire alarm system goes into alarm, you are certain that there is no fire and, despite your best efforts, you are not able to silence the alarm, there is further recourse. Perhaps some weird anomaly has affected the touchpad. As a last resort, you can open the control panel and locate the ac supply coming in. It will most likely consist of 12 AWG black, white, and green wires, connected to terminals. First, disconnect the dc power supply from the batteries. This can be done by disconnecting just one wire anywhere along the series string. Then, remembering that it is live, disconnect the black ac wire from the terminal. Use whatever protective measures you are comfortable with—rubber mat on the floor, insulated gloves, insulated tools, etc. Remember that after you pull the black wire, it is still live. Do not let it whip around and contact grounded metal inside the control panel or you will have a fireworks display. Put a wire nut on the end of the live wire to protect others from shock and to guard against a short. If there is still copper showing, add electrical tape.

Now that the control panel has been powered down, the alarm cannot sound. To reemphasize, it is necessary to ensure that the dc supply is disconnected first, before the ac supply is disconnected. When reconnecting, do the ac first. That way, the system is never on dc alone.

Notify the fire department, emergency call center, monitoring agency, and insurance company. Prior to starting the procedure, human fire patrols should be initiated throughout the area that is covered by this disabled fire alarm panel.

For a fire alarm technician, grounded zone conductors can make for the most difficult of all repairs. This fault can throw the fire alarm system into either the trouble or the alarm mode. Moreover, it can be intermittent, gone when the technician wants to work on it, reappearing in the night to awaken hotel guests. In the war on ground faults, you may have to open up multiple fronts. If the mini-megger procedure that we described earlier does not work, obtain a high-quality Fluke multimeter that will measure in the high-megohm ranges. This instrument will be costly, but very durable and aggressive on difficult jobs.

In an old legacy system, there are likely to be multiple ground faults, sometimes two or more in the same zone. This is particularly true when remodeling work is in progress and there are other trades banging around and drilling. The best troubleshooting procedure is to begin by disconnecting both wires of all zones simultaneously at the control panel. This will involve putting the system out of service, so the patrols and notifications discussed earlier will be needed. Starting at a low-ohm range and working up to your meter's limit of sensitivity, take readings between each wire and the grounded metal enclosure. (Verify the integrity of this ground before starting.)

This is like fishing for trout—there are times when they are not feeding and you will not get a bite. However, you know something is not right because there have been intermittent trouble and alarm states.

Have a helper monitor the meter at the control panel. It is a good idea to have leads with strong alligator clips so that any motion of the probes will not make the high-megohm readings jiggle all over the place. For this procedure to work, you need good cell phone or two-way radio communication.

First, remove the covers of all enclosures where there are zone terminations. Tighten and rearrange wire nuts and look for any wires that have been pinched under the covers or at connectors. If there are locknuts on the connectors, loosen them, rotate the connectors slightly, and then retighten the locknuts, all the while maintaining contact with the helper back at the fire alarm control panel. The idea is to watch for any change in the resistance in response to moving the wiring around.

If the intermittent was not active immediately prior to this procedure and is still not active, it is possible that the fault has been fixed and may never recur. In the same manner, another diagnostic test may be performed. If the metal raceway is surface mounted, tap the length of each run with a wooden broomstick, again with the helper monitoring the multimeter on a high-ohm setting. If there is a reaction, there is the fault. Damage can be due to moisture infiltration, vibration, high temperature (does the raceway cross a steam pipe?), stray nail or drill bit, or faulty initial installation. This cable would have to be pulled and replaced.

If the raceway is concealed behind wall or ceiling finish, the only thing you can do is tug on the cable at terminations and watch for resistance changes.

Drastic Measures

If none of these procedures work, you could do a wholesale replacement of cable, which is not as bad as it sounds, given the low cost of fire alarm cable and the ease of EMT pulls. Do not begin by pulling out the old cable. Instead, use it as a pull rope for the new installation. Even if this does not solve the problem, you know you have new wire for the zone circuit, and that is a plus.

Finally, it is possible that there is a fault inside the control panel. Look for any bad wires, test components, and check circuit boards. All it takes is a metal filing, bad solder joint, or a speck of conductive dirt to short out adjacent traces. Ordinarily, bad circuit cards show up as trouble events, but a certain fault could appear as a grounded conductor fault. Use your meter and the schematic to see if you can discover what is going on.

If the intermittent trouble or alarm condition persists, you have to face the possibility that in rewiring zones you introduced a new fault. Remember that just a slightly creased alarm cable can initiate a trouble state in the system. As stressed earlier, it is worth repeating that the control panel can differentiate between an open line and a normal condition because of the end-of-line resistor. It cannot distinguish a short between the two conductors and an alarm, so it will go into alarm if two conductors become shunted. This condition is quite rare because a bare copper situation for both conductors would have to occur at the same location.

If both wires contacted the raceway, even at opposite ends, that would constitute a short between conductors. Presumably, both faults would not occur at the same time. The first fault would make a trouble signal, grounded initiating device, or indicating appliance trouble report. For this reason, grounded trouble signals must not be tolerated. Once they proliferate, the troubleshooting task becomes more difficult and false alarms may ensue. The priority in all fire alarm work is to institute an aggressively proactive maintenance policy because human lives depend on the quality of your work. Any repairs become more difficult if deferred.

We have described some of the details of a fire alarm system. As the area covered becomes larger, the system becomes more complex and greater maintenance resources need to be allocated to keep it reliable and free of false alarms. Fire alarm systems are challenging, and definitely more difficult than other types of electrical troubleshooting and repair, with the possible exception of elevators. If you have nerves of steel and impeccable workmanship, this may be for you. Assuming you are an experienced electrician with good computer skills and a fundamental knowledge of electronics, some additional steps for entry into this field are as follows:

- Read and understand applicable codes and local statutes that pertain to fire alarm design and installation. These are not generally available on the Internet, but must be purchased or found in large technical school libraries.

- Carefully go through manufacturers' installation and user's manuals, which are available as free downloads from the Internet.

- If possible, work for a professional fire alarm installation and maintenance firm. This may coincide with obtaining fire alarm design and installation licensing in your state, if applicable. Absolutely no work should be performed off the books or under the table, as the stakes are too great. Licensing should be seen not as a chore, but as a learning opportunity.

An accomplished fire alarm professional occupies a position at the top of the field, as far as electrical work is concerned, and practitioners will never lack for gainful employment.

Troubleshooting Refrigeration Equipment

Commercial refrigeration equipment is everywhere. Cities and suburban areas have numerous restaurants, grocery stores, hotels, and so on, each with multiple refrigeration units of the walk-in variety, with remote compressors and condensers, as well as self-contained reach-ins. Pharmacies and healthcare facilities require freezers and refrigerators of many sizes and shapes, and some research and industrial complexes have huge refrigeration systems for super-conducting equipment to create zero resistance conditions at close to absolute zero. Then there are large office buildings with chillers, the capacity measured in tons. Additionally, air-conditioning is needed, even in winter, where there is a large concentration of electronic equipment, in telephone central office switch rooms and buildings that house large Internet provider servers and web hosts.

One thing all these cooling systems have in common is that when they go down, the heat is on in every meaning of the phrase. Time is of the essence to restore service because the contents will degrade quickly if the temperature rises above a critical point. (A medium-size walk-in freezer can contain upward of $15,000 worth of food).

It is assumed that a refrigeration technician will be a qualified electrician, with whatever licensing the jurisdiction requires. Additionally, in the United States an Environmental Protection Agency (EPA) administered license is required whenever a refrigerant circuit is opened. This licensing process focuses not on the competence of the technician to make repairs on refrigeration equipment but rather on the ability to add and remove refrigerant from the equipment without allowing any of it to be released into the atmosphere and find its way up to the ozone layer, which protects us from harmful ultraviolet radiation. One might ask: What possible harm, in a worldwide context, would the release of an ounce or so of this fluid do to a layer 10 to 30 miles above Earth's surface? If this quantity is multiplied by the millions of refrigeration compressors that are decommissioned, it is plain to see that protective measures need to be taken. That is what the EPA regulation is all about. Do not even think about breaking open a refrigeration circuit without the license. The fines are immense.

There are three levels of certification permitting technicians to work on different sizes of equipment. Acquiring the license is a good learning experience and lays part of the groundwork for gaining additional knowledge in the field. Then, one would do well to work with an experienced refrigeration technician while studying the technology.

Troubleshooting techniques we have been discussing are very applicable to refrigeration equipment because the constituent parts are connected in a linear fashion. Faults can be located using cause-and-effect reasoning and the half-splitting procedure.

The basic physics is interesting but a little counterintuitive. How can the application of energy cause there to be less energy (molecular motion = heat) at a certain location? Because of two heat-transfer events, energy is moved from one location, the refrigerated container, to another location, the outside world, where the heat is dissipated. Underlying this is a consequence that follows from Boyle's law. Robert Boyle (1627–1691) regarded acquisition of knowledge an end in itself. He formulated the law named after him, which may be stated as: "For an ideal gas kept at a fixed temperature, pressure and volume are inversely proportional." We shall see how refrigeration equipment is able to lower the temperature within a confined area by compressing a refrigerant outside of that area, allowing it to cool, and then moving it into the area to be cooled and decompressing it.

The principle parts of a refrigeration system are:

- The compressor
- The condenser
- The evaporator

The compressor is generally driven by an electric motor, most commonly 240-volt for small units and three-phase for larger ones. Years ago, there was a V-belt drive. This arrangement had the disadvantage that the compressor often developed a leak around the shaft, the seal having a finite life expectancy. To remedy this flaw, the hermetically sealed combination pump-compressor unit was developed and it is used today in most commercial-scale applications (see Fig. 13-1). Pump and motor are in a single enclosure and they run submerged in refrigerant. This arrangement is reliable and long lasting as long as

FIGURE 13-1 Hermetically sealed water-cooled compressor.

there is not a power surge from outside, loss of phase, or water contamination of the refrigerant, which can create an acidic mix that will attack the motor windings.

The compressor receives its supply of refrigerant via the low-pressure return line from the evaporator, which is inside the refrigerated box. As it reaches the compressor, the refrigerant is cool. The compressor increases the refrigerant's pressure, changing it from a gas to a liquid, so that it occupies a smaller volume. As part of the same process, it becomes much hotter. A good way to think of this is as a box containing many molecules moving rapidly and colliding with each other and the inside of the box, billiard ball style. If the box is made smaller, the molecules will begin to move more rapidly, the frequency and force of the collisions increasing. This is a consequence of the same amount of energy being squeezed into a smaller volume. The caloric value remains the same but the temperature increases.

Refrigerant Circuit

As it leaves the compressor, the refrigerant, now under increased pressure, is much hotter. If you were now to expand the contained refrigerant to its original precompressed volume, it would no longer be hot. It would return to its original temperature. Such a process could not be used for refrigeration. What makes the refrigeration process work is the next stage. The hot, compressed refrigerant enters the condenser, where its temperature is reduced to a level close to that of the air near the condenser.

> To lubricate the bearings of a furnace blower motor, it is often necessary to remove the squirrel cage fan from the motor shaft to access the motor, and this can be difficult. If available gear pullers will not fit, make a hardwood wedge to go behind the fan. Cut a slot in the middle of the wedge so that the two tapered fingers can go past the shaft. For stubborn cases, sand and wax the wedge and drive it in hard. Then use a wood dowel and hammer to strike the end of the shaft with the whole assembly elevated off the bench. Prepare the shaft in advance with penetrating oil if necessary. Disassemble the motor and oil both bearings. If, after reassembly, the motor appears to turn harder than ever, it is because one of the end housings has become bent and the bearings are out of line. It is not very difficult to straighten it and get everything going. When you are finished, you should be able to spin the shaft by hand and it will coast for a while.

The condenser has much the same form as an automobile radiator. The hot refrigerant enters the condenser where it flows through an assembly of small pipes that have cooling fins. The refrigerant cools down to much closer to room temperature, the process assisted by a fan that blows air across the cooling fins. (Some condensers are water cooled.) When the refrigerant leaves the condenser, it has become much cooler but it is still a liquid and still under pressure. It travels through a copper pipe and enters the refrigerator box through a grommeted hole. A necessary part of the picture is a small piece of hardware known as the diffuser valve, located in the refrigerant line adjacent to the box wall.

The diffuser valve consists of a small orifice, its exact size varying for different types of refrigerant. It sharply restricts the flow of refrigerant so that the pressure drops abruptly. Again, the refrigerant changes phase from a liquid to a gas. At this point, it is as if the box of billiard balls suddenly became larger. As the volume of the refrigerant per unit of mass

FIGURE 13-2   Five evaporators in a walk-in cooler.

increases, the temperature suddenly drops. Because it has lost heat in the condenser, it does not go to its precompressed room temperature. Instead, it becomes much colder. Inside the box, it travels to the evaporator (see Fig. 13-2). This again has the same general form as an automotive radiator. The now gaseous refrigerant travels through a system of small pipes with fins attached. A powerful fan blows air across the fins, and this is how the low temperature in the refrigerator box is maintained.

The unit can be either a refrigerator or a freezer depending on the size of the compressor, the size of the box, how well insulated it is, and the setting of the thermostat. The refrigerant leaves the box and travels through the return line back to the compressor where the cycle begins anew.

Refrigeration Repairs

The physical principles that make refrigeration possible are easy to understand, and most of the time troubleshooting and repair are straightforward. When a refrigeration unit fails in a commercial setting, for example a restaurant walk-in, the first responder is often the nearest electrician. This individual will look it over, get it running if possible, and make the decision to call in a licensed technician if it appears necessary to break open the refrigerant circuit. In this undertaking, the electrician will face two potentially disturbing realities. The first of these is that there is a lot of pressure to get the unit running immediately because the materials being refrigerated (food, pharmaceuticals, biological samples in a healthcare facility, etc.) will be compromised soon. As the temperature rises, the situation becomes increasingly dramatic. Second, if you succumb to the temptation to cop a scenario and announce that the compressor needs to be replaced, and if this is done at great expense, the system still does not work, and it turns out to be something simple, this could be damaging.

Therefore, if it seems that you might be doing this sort of work, some careful study and contingency planning are in order.

When you are called in to look at a refrigeration system that is not working, most likely the complaint is that the temperature is too high. It is either not cooling enough or not cooling at all. Occasionally, the temperature is too low; for example, the lettuce is freezing in a produce walk-in. If this is the case, or if the temperature is a few degrees too high, most likely the remedy is to adjust the thermostat. Just tweak it a slight amount. Then come back in a couple of hours and check it again. This is the most common refrigeration repair and it is usually successful.

Next in order of frequency is when a refrigeration unit is in need of defrosting. The initial complaint will be that the unit is running warm. All refrigeration equipment should have a means for checking the temperature. This can be a sensor with a digital readout on the outside, or it can be a thermometer inside. The users should be encouraged to check the temperature regularly in order to spot trouble before it becomes severe. Icing will start gradually and become worse. Eventually, the temperature inside will not be much different from the outside. The cause is that the outside of the evaporator has become coated with ice and even though the fan is blowing air across the ice, there will be very poor thermal transfer. In severe cases, the ice will interfere with the fan blade, causing the motor to stall out and trip the overcurrent protection or destroy the motor.

There are several possible causes of icing. It happens more in summer when the weather has been hot and humid. One way or another, moisture-laden air enters the box. It is a principle of physics that water vapor migrates from a warmer to a colder region. In a refrigeration unit, moisture collects on the outside of the evaporator, forming a layer of ice. A contributing factor may be that workers are going in and out of the walk-in, leaving the door open or ajar when they are inside. They should be asked to plan their tasks so that trips inside are kept to a minimum, especially on muggy days.

Another cause of frost build-up on the evaporator is a door that is out of adjustment. Moist air can enter around the edges of the door, and if it is a freezer, ice will build up here, progressively preventing the door from closing, making the gap larger and accelerating the process. Watch out for a worn door gasket. It allows moist air to enter and it should be repaired or replaced.

If the box is not urgently needed, the remedy is simple. Power down the compressor for a few hours and let the ice melt. Prop the door open during this period. Placement of a fan or portable heater near the evaporator will speed things up.

If the unit is in use and urgently needed, the ice will have to be removed manually. When doing so, always power down the compressor first. Otherwise, you will be battling the cold.

Use a small heat gun or a propane torch to melt out the block of ice. As soon as chunks of ice become loose, remove them to speed up the process, taking care not to pull out any wires that may be embedded. If using a propane torch, keep it moving. Too much heat will damage the aluminum fins and nearby components. It is amazing how long it takes to remove a small block of ice.

Some preventive measures may be taken. One of these is a heating element that is made to go on an evaporator. It is controlled by a timer and goes on periodically to control ice formation. Another approach is to install a 24-hour clock timer in the compressor circuit so that the refrigeration equipment goes into a defrost cycle periodically. These cycles can be set to coincide with times of decreased usage, such as after midnight.

Wine coolers, ice machines, and drinking fountains work on these same principles. Depending on the use, different temperatures should be selected by setting the thermostat appropriately. Freezers should be very cold, the temperature as low as −20°F. The colder it is, the better food is preserved. In addition, this buys time in case of malfunction. An exception is an ice cream freezer, which should be set at around 20°F so that the ice cream can be easily scooped. Walk-in and reach-in coolers should be set at around 40°F depending on the type of food (produce, soups, etc.) being stored. The head chef will have definite ideas on this. Food safety and preventing freezer burn are big topics that need to be kept in mind in the maintenance of refrigeration equipment.

The two failure modes mentioned previously—thermostat misadjustment and icing—account for a good proportion of refrigeration problems, and they are easily solved, at least on a temporary basis. Long-term solutions should be sought when these problems are found to be chronic. The high-pressure line should be very hot where it comes out of the compressor, much cooler after the condenser, and cooler still due to heat loss in the piping as it approaches the box, depending on the length of the line.

The compressor will make a loud characteristic noise if it is running, which almost resembles a knocking sound. In addition, it will be quite hot to the touch. These are both natural consequences of the compression that is taking place.

Compressor Fault Modes

If the compressor is running but there is no cooling action, there are a number of protective functions that may kick in, eventually cutting off the electrical supply to the compressor. One of these is over-temperature and another is blockage of refrigerant flow. If the compressor has not shut down, one of the sensors may be bad or a relay could be stuck. When a relay is stuck, lightly tapping on it can restore action, but it is a sign that the relay needs to be serviced or replaced.

The system can be checked, if the compressor is running, by feeling the refrigerant piping at various points. The high-pressure line may be identified by the fact that it is smaller than the return line, which operates at lower pressure and conveys refrigerant in its gaseous form. The high-pressure line should be very hot where it comes out of the compressor, a little cooler due to heat loss in the piping as it approaches the box, depending on the length of the line. Immediately after the diffuser valve, the pipe will be much colder up to the evaporator. After the evaporator, it will be not so cold, and it will continue to warm as it travels to the compressor. Performing this check when the compressor is running will tell you a lot about the status of the system. Check the obvious things. The fans should be running, the condenser and evaporator fins not obstructed with dirt, the belts not slipping, the motors not overheating, etc. Clamp-on ammeter readings can be taken at the compressor and the two fan motors. These motors, if they have provision for external lubrication, should be oiled lightly at the prescribed intervals.

Another useful observation is to look at the sight gauge, which allows you to monitor the amount of refrigerant in the system. The only problem is that if there is no refrigerant or too much refrigerant, then both conditions will appear the same. It is difficult to discriminate between pure air and pure refrigerant. The normal condition is to see just a few occasional bubbles, moving along with the flow of the refrigerant away from the compressor and toward the diffuser valve. If there is too much refrigerant, it can cause the compressor to lock up and not run.

In either of these situations, it will be necessary to call in a licensed refrigeration technician to add or remove refrigerant. It is not unusual to have to add refrigerant occasionally, but if the problem recurs, there is a leak in the system. These leaks may be quite difficult to pin down, especially if the compressor is remote from the box, if all or part of the piping is concealed behind wall, floor, or ceiling finishes, or if other piping or ducts are in the way. There are diagnostic tools such as the refrigerant sniffer that makes an audible indication when refrigerant, in gas form, is detected in the air, even in very small quantities.

> The National Electric Code requires that emergency lights are to be tested periodically and records kept. The time interval is not specified. Pressing the test button interrupts ac power so that the lights should go on. If they do not, put in new batteries. Then the lights will go on when you push the test button unless the new batteries are discharged, the bulbs are out, or there is a break in the wiring. If the new batteries are not charged the next time the unit is tested, the charger or relay may be bad. You do not need to buy a new box. Just change the circuit board.

If the compressor will not start, any number of conditions could be the cause. See if there is voltage at the disconnect output and at the motor relay input. If there is voltage, but the relay has not pulled in, check for voltage at the control terminals. Any one of the control devices can interrupt operation. The quick and dirty way to isolate the problem is to shunt out the devices individually using an insulated jumper. The problems with this method are danger of shock (if you slip), danger of arc flash (if you slip or there is a short downstream), and damage to components. It is better to use a meter, taking live voltage measurements and powered-down resistance measurements.

If the compressor runs continuously, there can be a design problem. This is where the user interview is useful. If the refrigerated box has been increased in size, the compressor may run continuously. Other possibilities are that there is insufficient refrigerant in the system, the evaporator is iced or dirty, the evaporator fan motor is not working, or the thermostat is defective.

If the compressor will not start, but makes a 60-cycle hum, the cause can be low voltage in the power supply, bad capacitor or start relay, shorted or grounded windings in the motor, or a seized compressor.

If the compressor will not start and it makes no sound, it can be an open motor winding, motor protector, or circuit breaker. Also, look for a defective temperature or pressure control.

A good preventive maintenance program is essential. Air-cooled condensers (see Fig. 13-3) collect dust, and this interferes with their ability to get rid of heat, making the compressor work long hours. The condenser core and in fact the entire area should be vacuumed out as needed. If the compressor room is remote from the refrigerated box, especially if there are several compressors grouped, ventilation is a big concern. Moreover, the condenser should be located away from any heat source. If room ventilation is dependent upon a fan, it should be checked frequently, lubricated as needed, and it should be on a dedicated circuit. These problems are eliminated by having water-cooled condensers because there are no moving parts and less tendency to accumulate airborne dust. Remember that the objective is to dissipate heat. The more efficiently this takes place, the less the compressor has to work, meaning lower energy costs and a longer time until the next repair. If air from the compressor room is exhausted to the outdoors, it means that an equal amount of air has to

Figure 13-3   Three air-cooled compressors in a compressor room.

enter the room to replace it, and that is how the dust gets in. Consider installing a furnace-type air filter so that air entering the room is clean.

Dust and dirt are much less of a problem at the evaporator. Here the problem is ice. It has much the same effect, interfering with heat transfer and making the system work longer, so good maintenance is called for. V-belts on the evaporator and condenser should be checked regularly for wear and proper tension, and motors lubricated. Keep a watch on the sight gauge and asked the licensed refrigeration technician to add refrigerant when necessary.

A problem with hermetically sealed compressors occurs when the refrigerant becomes contaminated with water. Just a small amount of moisture in the fluid will make an acidic mix that will etch through the insulating coating on the motor windings. This will ground them out, essentially ruining the hermetically sealed motor/compressor. For this reason, a water separator is put in the line. Water enters the refrigerant circuit when it is opened for servicing. Air will enter, and since there is a certain amount of moisture in ordinary environmental air, it enters the circuit and combines with the refrigerant.

In designing a refrigeration installation, consider the following:

- Just as heat must be permitted to dissipate from the area of the condenser, it should be prevented from entering the box. If there are two walk-ins, they should be located adjacent to one another with a common wall. When there are four walk-ins, they should be grouped in a cube configuration. Self-contained reach-ins should be located away from sources of heat.

- Evaporators in a walk-in require condensate trays so that the water that accumulates can be captured and drained. A sheet-metal tray can be fabricated with copper tubing exiting the box so that water can be disposed of. The floor of a walk-in is usually concrete or tile and it should have a floor drain in the center with the floor sloping toward it. This has to be factored in at the design stage.

- The entire walk-in can be built from locally available materials, with the exception of door hardware, compressor, condenser, evaporator, and associated hardware. Inside surfaces should be waterproof and walls and ceiling well insulated.

Back to NEC

Electrical requirements for refrigeration are laid out in Article 440, Air-Conditioning and Refrigeration Equipment. This article follows Article 430, Motors, Motor Circuits and Controllers. The requirements for refrigeration equipment that includes electrical motors differ somewhat from general electrical requirements, especially concerning sizing out the supply circuit. As we have seen in Chap. 2, most electrical motor supply circuits are sized out not by referring to the current rating on the nameplate of the motor, but by taking the horsepower value off the nameplate and referring to Tables 430.247, 248, 249, and 250. These tables give full-load current values for dc, single-phase, two-phase (rare), and three-phase motors at various voltages. The tables are found near the end of Article 430 and this is the way to size out motor circuits for most applications.

Air-conditioning and refrigeration equipment circuits are sized out in a different manner. Article 440 specifically applies to hermetically sealed refrigeration compressor/motors. It is applicable concerning sizing out disconnecting means, controllers, and single or group installations, as well as sizing out their conductors.

What is different for hermetically sealed refrigerant compressor/motor combinations is that they have on the nameplate a value that is called the "branch-circuit selection current." To emphasize, this value is to be used instead of the rated-load current to determine the size of the disconnecting means, the controller, the motor branch-circuit conductors, and the overcurrent protective devices for the branch-circuit conductors and the motor. It is to be noted that the value of branch-circuit selection current is always greater than the rated-load current that is shown on the nameplate. If this provision is not understood and followed, the circuit could be undersized.

For an electrician or technician who is troubleshooting or repairing refrigeration equipment, a priority should be to verify that the original installation, especially from an electrical point of view, is fully compliant. If new supply wiring and overcurrent protection has to be installed, that is part of the cost of doing business.

Disconnecting means are rated at 115 percent of the refrigeration equipment nameplate rated-load current or branch-circuit selection current, whichever is greater.

For cord-connected equipment, such as room air-conditioners, household refrigerators and freezers, and similar appliances, the plug and receptacle will suffice as the disconnecting means.

The disconnecting means must be in sight of equipment, so that in repair operations, the technician can be certain that the equipment will remain powered down. V-belt and fan-blade injuries can be severe.

We have mentioned a few of the NEC mandates that pertain to refrigeration equipment just to give an idea of the type of requirements that have to be observed. In contemplating this type of work, one of the responsibilities is to review thoroughly the regulatory material to make sure that costly and hazardous mistakes or omissions do not occur.

Healthcare Facilities

Healthcare facility wiring is exacting and requires a great amount of knowledge and expertise. The systems are complex, bringing together many related elements. There are fire alarm systems, elevators, redundant grounding concepts, alternate power sources, places of assembly, hazardous locations, and a lot of data cabling, all combined in a single building or set of buildings. Always, the objective is to protect a group of individuals, the patients, from further injury. They are vulnerable to hazards, particularly electrical, and for this reason, care must be taken to create healthcare environments that have built-in safeguards.

One of the reasons that patients are vulnerable is that, often in a weakened state, they are by necessity required to undergo intrusive medical procedures that involve conductive metal instruments placed in the bloodstream. Blood is much more electrically conductive than water, and the veins and arteries are pipes connected to the heart. Its operation is disrupted by a slight amount of electrical current, so measures must be in place to ensure that these instruments either are insulated or remain at ground potential. In addition, conductive surfaces within reach of the patient or a staff member who may touch the patient must be intentionally and reliably grounded. For many patients, even a mild shock could be fatal.

There are other issues as well. A certain portion of the patient population at a given time may be on life support. These patients' lives depend upon an electrical supply that cannot be interrupted. As we all know, utilities endeavor to keep their lines energized at all times, but severe weather or mechanical failure can stop the flow of electrons. Moreover, it is not solely a matter of life support. The efficient operation of the entire healthcare facility depends upon a reliable supply of electricity, and the quality of patient care can be impacted in many ways. For these reasons, healthcare facilities are required to have backup power, and the specifications are explicit. The electrical load is divided into sectors, the exact number and nature of them depending upon the usage and size of the occupancy. To design and build or to competently maintain, troubleshoot, and repair electrical equipment in these buildings, it is necessary to understand the infrastructure requirements. For a start, we will look at the National Electrical Code (NEC), and see how it defines the various occupancies and electrical equipment within them. (Two other documents are relevant: NFPA 99-2005, Standard for Healthcare Facilities, and NFPA 101-2009, Life Safety Code. Both of these works are referenced in NEC 2011, and the careful electrician will want to consult them before undertaking any significant healthcare facility work.)

Know Your Definitions

To get started, it is suggested to assimilate and keep the following definitions in the foreground:

Alternate Power Source: A motor-driven generator that will provide power during a utility outage. Sometimes backup batteries perform this function. In some cases, the generator set is the primary power source and the utility is the backup.

Ambulatory Healthcare Occupancy: An outpatient building for four or more patients that provides treatment or anesthesia that renders them incapable of taking action for self-preservation under emergency conditions without assistance of others, or provides emergency or urgent care for patients who, due to the nature of their injury or illness, are incapable of taking action for self-preservation under emergency conditions without the assistance of others.

Anesthetizing Location: Any area of a facility where a flammable or nonflammable inhalation anesthetic may be used. This includes their use for relative anesthesia, where the patient is not made entirely unconscious. Flammable inhalation anesthetics are no longer generally used in the United States, but the Code still refers to them.

Critical Branch: A subsystem of the emergency system that consists of feeders and branch circuits supplying electricity for lighting, special power circuits, and selected receptacles serving areas and functions related to patient care and that are connected to alternate power sources by one or more transfer switches during interruption of the normal power source.

Critical Care Areas: Those special care units, intensive care units, coronary care units, angiography laboratories, cardiac catheterization laboratories, delivery rooms, operating rooms, and similar areas in which patients are intended to be subjected to invasive procedures and connected to line-operated electromedical devices.

Electrical Life-Support Equipment: Electrically powered equipment whose continuous operation is necessary to maintain a patient's life.

Emergency System: A system of circuits and equipment intended to supply alternate power to a limited number of prescribed functions vital to the protection of life and safety.

Equipment System: A system of circuits and equipment arranged for delayed, automatic, or manual connection to the alternate power source that serves primarily three-phase power equipment.

Essential Electrical System: A system comprised of alternate sources of power and all connected distribution systems and ancillary equipment, designed to ensure continuity of electrical power to designated areas and functions of a healthcare facility during disruption of normal power sources, and also to minimize disruption within the internal wiring system.

Exposed Conductive Surfaces: Those surfaces that are capable of carrying electric current and that are unprotected, unenclosed, or unguarded, permitting personal contact. Paint, anodizing, and similar coatings are not considered suitable insulation.

General Care Areas: Patient bedrooms, examining rooms, treatment rooms, clinics, and similar areas in which it is intended that the patient will come in contact with ordinary appliances such as a nurse call system, electric beds, examining lamps, telephones, and entertainment devices.

Hazard Current: For a given set of connections in an isolated power system, the total current that would flow through a low impedance if it were connected between either isolated connector and ground. The fault hazard current is the hazard current of a given isolated system with all devices connected except the line isolation monitor. The monitor hazard current is the hazard current of the line isolation monitor alone. The total hazard current is the hazard current of a given isolated system with all devices, including the line isolation monitor, connected.

Healthcare Facilities: Building or portions of buildings in which medical, dental, psychiatric, nursing, obstetrical, or surgical care is provided. They include but are not limited to hospitals, nursing homes, limited care facilities, clinics, medical and dental offices, and ambulatory care centers, whether permanent or movable.

Hospital: A building or portion thereof used for 24-hour medical, psychiatric, obstetric, or surgical care for four or more inpatients.

Isolated Power System: A system comprising an isolating transformer or its equivalent, a line isolation monitor, and its ungrounded circuit conductors.

Isolation Transformer: A transformer of the multiple winding type, with the primary and secondary windings physically separated, which inductively couples its secondary windings to circuit conductors connected to the primary windings.

Life Safety Branch: A subsystem of the emergency system consisting of feeders and branch circuits intended to provide adequate power needs to ensure safety to patients and personnel, and that are automatically connected to alternate power sources during interruption of the normal power source.

Limited Care Facility: A building or portion thereof used for 24-hour housing of four or more persons who are incapable of self-preservation because of age, physical limitation due to accident or illness, or limitations such as mental retardation/developmental disability, mental illness, or chemical dependency.

Line Isolation Monitor: A test instrument designed to continually check the balanced and unbalanced impedance from each line of an isolated circuit to ground and equipped with a built-in test circuit to exercise the alarm without adding to the leakage current hazard.

Nurses Stations: Areas intended to provide a center of nursing activity for a group of nurses serving bed patients, where the patient calls are received, nurses are dispatched, nurses' notes are written, inpatient charts are prepared, and medications are prepared for distribution to patients.

Nursing Home: A building or portion of a building used for 24-hour housing and nursing care of four or more persons who, because of physical or mental incapacity, might be unable to provide for their own needs and safety without assistance of others.

Patient Bed Location: The location of a patient sleeping bed, or the bed or procedure table of a critical care area.

Patient Care Area: Any portion of a healthcare facility wherein patients are intended to be examined or treated. Areas of a healthcare facility in which patient care is administered are classified as general care areas or critical care areas. Business offices, corridors, lounges, day rooms, dining rooms, or similar areas are not patient care areas.

Patient Care Vicinity: In an area in which patients are normally cared for, the patient care vicinity is the space with surfaces likely to be contacted by the patient or an attendant who can touch the patient. Typically in a patient room, this encloses a space within the room not less than 6 ft beyond the bed in its nominal location and extending not less than 7½ ft above the floor.

Patient Equipment Grounding Point: A jack or terminal that serves as the collection point for redundant grounding of electrical appliances serving a patient care vicinity or for grounding other items to eliminate electromagnetic interference problems.

Psychiatric Hospital: A building used exclusively for 24-hour psychiatric care of four or more patients.

Reference Grounding Point: The ground bus of the panelboard or isolated power system panel supplying the patient care area.

Selected Receptacles: A minimum number of electrical receptacles to accommodate appliances ordinarily required for local tasks or likely to be used in patient care emergencies.

Wet Procedure Locations: Those spaces within patient care areas where a procedure is performed and that are normally subject to wet conditions while patients are present. These include standing fluids on the floor or drenching of the work area, of which either procedure is intimate to the patient or staff. Routine housekeeping procedures and spillage of liquids do not define a wet procedure location, nor do bathrooms, lavatories, or sink areas.

These definitions provide a useful overview of the range of healthcare facility types, as well as some of the circuits and equipment encountered in a healthcare facility. A veterinary office or animal clinic is not considered a healthcare facility. It is possible for a healthcare facility to be part of a larger building, such as a clinic inside an office building. These healthcare facility variants are referenced in other parts of Article 517, where wiring methods are discussed, as well as the requirements for alternate power sources and how they are connected to specific loads.

To really get on top of this, you need a good mental picture of how the loads subdivide and connect by means of panels and transfer switches to the normal power source and the backup source.

Part II of Article 517 concerns wiring and protection in healthcare facilities. It does not apply to business offices, corridors, waiting rooms, and the like in clinics, medical and dental offices, and outpatient facilities; nor does it apply to areas in nursing homes and limited care facilities used exclusively as patient sleeping rooms.

When You Need Redundant Grounding

In Section 517.13, Grounding of Receptacles and Fixed Electrical Equipment in Patient Care Areas, the principle protective measure that guards against electric shock for patients is outlined. It provides that all branch circuits serving patient care areas be provided with an effective ground-fault current path by installation in a metal raceway system or a cable having a metallic armor or sheath assembly. The metal raceway system or metallic cable armor or sheath assembly must itself qualify as an equipment-grounding conductor. This requirement is quite broad. It includes patient care areas in nonhospital settings, such as nursing homes, clinics, and medical and dental offices. It is not limited to patient rooms, but also includes examining rooms, therapy areas, recreational areas, and the like.

In addition to the ground path described previously, all wiring in these areas is required to have an insulated grounding conductor. This is what redundant grounding means. This wire must attach to the grounding terminals of all receptacles, metal boxes, and enclosures containing receptacles and all noncurrent-carrying conductive surfaces of fixed electrical equipment likely to become energized that are subject to personal contact, operating at over 100 volts. This grounding conductor must be an insulated copper wire, solid or stranded, that is installed with the branch-circuit conductors in one of the wiring methods that qualifies as an equipment grounding conductor as described previously.

Together, these two ground paths comprise the redundant grounding arrangement that is required for patient care areas, broadly defined. It is not necessary to run the redundant grounding wire upstream from a load center (that has overcurrent devices) to the entrance panel. In addition, it is not necessary to connect the insulated redundant grounding conductor to metal faceplates, which are bonded to the grounded metal strap of the device. A further exception exempts luminaires (light fixtures) 7½ ft above the floor and switches located outside the patient care vicinity.

Section 517.14, Panelboard Bonding, provides that the equipment-grounding terminal buses of the normal and essential branch-circuit panelboards serving the same individual patient care vicinity be connected together by means of an insulated continuous copper conductor not smaller than 10 AWG. Where two or more panelboards serving the same individual patient care vicinity are served from separate transfer switches on the emergency system, the equipment-grounding terminal buses of those panelboards are to be connected together with an insulated continuous copper conductor not smaller than 10 AWG. This conductor is permitted to be broken in order to terminate on the equipment-grounding terminal bus in each panelboard.

Section 517.18, General Care Areas, provides that each bed location be supplied by at least two branch circuits, one from the emergency system and one from the normal system. All branch circuits from the normal system must originate in the same panelboard. The purpose of the first of these requirements is to ensure that there will be continuity of power for life-support, and diagnostic and therapeutic equipment. The purpose of the second of these requirements is that in the event of a sudden unintended fault that affects a branch circuit terminated at one of the panels, there would not be a dangerous voltage differential between the separately connected equipment grounding conductors. It is further stipulated that the branch circuit serving patient bed locations is not to be part of a multiwire branch circuit.

Exceptions provide that these requirements are not applicable to the following:

- Branch circuits serving only special-purpose outlets or receptacles, such as portable X-ray outlets.
- Patient bed locations in clinics, medical and dental offices, outpatient facilities, psychiatric, substance abuse, and rehabilitation hospitals, and sleeping rooms of nursing homes and limited care facilities.

How Many Receptacles?

Each patient bed location is to be provided with a minimum of four receptacles. They may be single, duplex, quad, or any combination, as long as there is a minimum of four at each general care bed location. All receptacles, even if there are more than the minimum four required, are to be listed and identified as hospital grade. They are identified by having a green dot on the face of the receptacle. As noted previously, the grounding terminal must be connected to the insulated copper redundant equipment grounding conductor.

Section 517.19, Critical Care Areas, parallels the preceding section on general care areas, but some of the requirements are more stringent. It lays out the electrical circuit, receptacle, grounding, and bonding requirements for these areas. Recall that critical care areas are where patients are intended to be subjected to invasive procedures and to be connected to line-operated, electromedical devices. An example is the intensive care unit, but the category also includes several less demanding locations.

Within a critical care area, each patient bed location is to be supplied by at least two branch circuits. There must be one or more from the emergency system and one or more from the normal system. At least one branch circuit from the emergency system must supply one or more outlets only at that bed location. All branch circuits from the normal system must be fed from a single panelboard. Emergency system receptacles are to be identified and must indicate the panelboard and circuit number supplying them. Once again, no multiwire circuits are permitted.

An exception provides that critical care locations served from two separate transfer switches not be required to have circuits from the normal system.

It is common practice to color-code the receptacles supplied by the emergency system. Red is customarily used for both the receptacle and the faceplate. Additionally, the receptacle will have the green dot indicating that it is hospital grade.

Each patient bed location must be provided with a minimum of six receptacles, as opposed to the four required for patient bed locations within general care areas. Here again, however, they may be single, duplex, quad, or any combination.

The metal raceway that qualifies as an equipment grounding conductor must attach to enclosures and equipment such as panelboards and switchboards in such a way as to ensure that there is no loss of ground continuity. One of the following bonding means must be used at each termination or junction point:

- A grounding bushing and a continuous copper bonding jumper connected to the junction enclosure or the ground bus of the panel.

- Connection of feeder raceways to threaded hubs or bosses on terminating enclosures.

- Other approved devices such as bonding-type locknuts or bushings.

Additional optional protective techniques are permitted for critical care areas. Isolated power systems are used in operating rooms and other sensitive locations. Power for loads is conveyed via branch circuits connected to the secondary of an isolation transformer. For single-phase systems, there are two conductors, both isolated from ground. Isolated conductor no. 1 is identified by orange insulation with at least one distinctive colored stripe other than white, green, or gray along the entire length. Isolated conductor no. 2 is identified by brown insulation with at least one distinctive colored stripe other than white, green, or gray along the entire length.

The isolating transformer or other isolated power circuit equipment is not to be located within a hazardous anesthetizing location. The secondary conductors that run into this classified location must be installed in accordance with NEC Article 501, Class I (Hazardous) Locations.

In addition to overcurrent protection, the isolated power system is to have a line isolation monitor. It has a green signal lamp that indicates that both lines are isolated from ground. A red signal lamp and audible alarm indicates that there is a leakage current to ground from either line greater than 5 mA. An ammeter is mounted on the line isolation monitor with the fault level at the center of the scale.

The equipment-grounding conductor may be run outside the raceway that contains the ungrounded conductors, in order to maintain the required high level of impedance between each of the lines and ground.

Wet procedure location patient care areas, defined previously, require special protective measures, and they fall into one of these categories:

- A power distribution system that inherently limits the possible ground-fault current due to a first fault to a low value, without interrupting the power supply.

- A power distribution system in which the power supply is interrupted if the ground fault current exceeds 6 mA.

It is to be noted that GFCI protection is not required to be installed in critical care areas where the toilet and basin are installed within the patient room. The GFCI protection is optional, neither required nor prohibited, and the judgment has to be made between protection from shock and reliability because there may be an issue of nuisance tripping.

Part III, Essential Electrical System, specifies the loads and connection procedures for that part of a healthcare facility's electrical infrastructure that is necessary for life safety and orderly cessation of procedures during the time normal electrical service is interrupted.

The entire electrical structure of a healthcare facility is composed of two parts—the nonessential electrical system and the essential electrical service. As its name implies, the essential electrical system is to receive power from an alternate power source during the period of time when the normal power source is disrupted.

Section 517.30 defines and lays out the requirements for essential electrical systems for hospitals. It is stated that essential electrical systems for hospitals are to be made up of two separate systems capable of supplying a limited amount of lighting and power considered essential for life safety and effective hospital operation during the time the normal electrical service is interrupted for any reason. The two systems are the emergency system and the equipment system.

The emergency system is made up of the life safety branch and the critical branch. The equipment system is not subdivided further. It supplies large units of electrical equipment necessary for patient care and hospital operation. This is primarily three-phase, much of it motor-driven.

Nonessential loads are connected directly to the normal source. When that power is interrupted, the nonessential loads go down and are not reactivated until normal power is restored. No transfer switch is required.

When the normal power source is operating, the essential electrical system is connected to it through one or more transfer switches, which by necessity are automatic rather than manual.

If a hospital has a maximum demand of 150 kW on the essential electrical system, that system may be switched between the two sources by means of a single transfer switch. For larger hospitals, there will be three automatic transfer switches, one for the critical branch, one for the life safety branch, and one for the equipment system. The transfer switch for the equipment system is of the delayed type to split the heavy inrush current.

If there are optional loads not required by the Code, they are not to be transferred if that would overload the generating equipment. Moreover, they are to be automatically shed if the generating equipment becomes overloaded.

The life safety branch and the critical branch of the emergency system are to be kept entirely independent of all other wiring and equipment and are not to enter the same raceways, boxes, or cabinets with each other or with other wiring.

Life safety and critical branch wiring are permitted to occupy the same raceways, boxes, or cabinets of other circuits if the wiring:

- Is in transfer equipment enclosures.
- Is in exit or emergency luminaires supplied from two sources.
- Is in a common junction box attached to exit or emergency luminaires supplied from two sources.
- Is for two or more emergency circuits supplied from the same branch and same transfer switch.

In contrast to the above requirements that pertain to the critical branch and the life safety branch, the wiring of the equipment system is permitted to occupy the same raceways, boxes, and cabinets of circuits that are not part of the emergency system.

Systems within Systems

The essential electrical system is made up of the equipment system and the emergency system, which is further subdivided into the critical branch and the life safety branch. Within 10 seconds after interruption of normal power, the two parts of the emergency system are to be connected to the alternate power source and simultaneously disconnected from the normal power source.

The life safety branch is to supply power to the following loads and no other loads:

- Illumination of means of egress, that is, corridors, passageways, stairways and landings at exit doors, and all necessary ways of approach to exits.
- Exit signs.
- Alarm and alerting systems including fire alarms and alarms required for systems used for the piping of nonflammable medical gases.
- Communications systems.
- At generator and transfer switch locations, battery chargers for battery-powered lighting units, and receptacles at the generator and transfer switch.
- Generator set accessories to facilitate operations.
- Elevator cab lighting, control, and communication and signal systems.
- Automatic doors used for building egress.

The critical branch of the emergency system is to provide power for lighting, fixed equipment, receptacles, and power circuits in these areas and functions related to patient care:

- Critical areas that use anesthetizing gases.
- Isolated power systems.
- Lighting and receptacles in patient care areas for infant nurseries, medication preparation areas, pharmacy dispensing areas, acute nursing areas, psychiatric bed areas, ward treatment rooms, and nurses' stations.
- Specialized patient care lighting and receptacles.
- Nurse call systems.
- Blood, bone, and tissue banks.
- Telephone equipment rooms and closets.
- Lighting, receptacles, and power for general care beds, angiographic labs, cardiac catheterization labs, coronary care units, hemodialysis rooms, emergency rooms, physiology labs, intensive care units, and postoperative recovery rooms.
- Additional lighting, receptacles, and power as needed.

The critical branch may be further subdivided.

The equipment system is to connect the following equipment to the alternate source after a suitable delay:

- Central suction systems for surgical and medical functions. (These could alternately be on the critical branch.)
- Sump pumps.
- Compressed air systems, alternatively permitted on the critical branch.
- Smoke control and stair pressurization.
- Kitchen hood ventilation and exhaust.
- Ventilation for airborne infectious isolation rooms, protective environment rooms, exhaust fans for laboratory fume hoods, nuclear medicine areas where radioactive material is used, ethylene oxide, and anesthetic evacuation. These may be placed on the critical branch if delayed automatic connection is objectionable.
- Supply, return, and exhaust ventilating systems for operating and delivery rooms.

Section 517.35, Sources of Power, provides that the essential electrical systems of a healthcare facility have at least two independent sources of electrical power. They are the normal source, which ordinarily supplies the entire occupancy, and at least one alternate source for connection when the normal source is disrupted. The alternate source is to take one of the following forms:

- One or more generators, located on the premises and driven by a prime mover, generally a diesel engine.
- A second generating unit where the normal source consists of a generating unit located on the premises.

- An external utility service when the normal source is a generating unit located on the premises.
- A battery system located on the premises.

Components of the essential electrical system must be carefully located to minimize interruptions caused by natural forces such as storms, floods, earthquakes, or hazards created by adjoining structures or activities. Consideration must also be given to the possible interruption of normal electrical service resulting from similar causes as well as possible disruption of normal electrical service due to internal wiring and equipment failures.

Nursing homes and limited care facilities also have the 150 kilovolt-amperes cutoff point concerning the minimum number of transfer switches required. The difference, however, is that an equipment system is not required. For the smaller occupancy, one transfer switch is needed and for the larger occupancy, two transfer switches are needed.

Part IV, Inhalation Anesthetizing Locations, provides that where flammable anesthetics are employed, the entire area be considered a Class I, Division 1 area up to a level 5 ft above the floor. This is largely a moot point in the United States, where flammable anesthetics are no longer used. The requirement remains in the Code because it has been enacted in some jurisdictions outside the United States where flammable anesthetic may still be used. Furthermore, use of flammable anesthetics could return, so the perspective this material provides could be valuable in the future.

In other-than-hazardous anesthetizing locations, where the anesthetics that are used are nonflammable, there are some requirements related to wiring methods. In these locations, conductors are to be installed in a metal raceway system or cable assembly, which must qualify as an equipment-grounding means. These locations are, generally, operating rooms, and it is mandated that one or more battery-powered lights be provided. The reason is that in a surgical procedure during the 10-second period between interruption of normal power and startup of the alternate source, the period of darkness could disrupt the procedure at a critical juncture. The battery-powered backup lights come on instantaneously, like emergency lights in a public place, so there is no period of darkness. Their effectiveness depends upon periodic testing to ensure that the batteries and charger are operational.

In any anesthetizing area, whether or not classified as hazardous, all metal raceways and all normally noncurrent-carrying conductive portions of fixed electrical equipment are to be connected to an equipment-grounding conductor.

Isolated power system equipment is permitted to be located in an anesthetizing location, provided it is installed above any hazardous (classified) location or in an other-than-hazardous location.

Part VI, Communications, Signaling Systems, Data Systems and Systems Less than 120 Volts, Nominal, contains requirements for these types of wiring within patient care areas. The principle mandate is that equivalent insulation and isolation to that required for the electrical distribution systems in patient care areas are to be provided for communications, signaling systems, data system circuits, fire alarm systems, and systems less than 120 volts nominal.

Class 2 and Class 3 signaling and communications systems and power-limited fire alarm systems are not required to comply with the grounding requirements of Section 517.13, to comply with the mechanical protection requirements in Article 517, or to be enclosed in raceways, unless specified elsewhere in the Code.

The foregoing is intended as an overview of the principle NEC requirements for healthcare facilities. It is beyond the scope of this volume to provide complete coverage of healthcare facility requirements in that Code, not to mention NFPA 99, Standard for Health Care Facilities, and NFPA 101, Life Safety Code. To become adept in healthcare wiring design and installation, one would have to study the entire NEC and the other documents intensively and to accumulate on-the-job experience working for a firm that performs this type of construction. Additionally, experience in related fields such as fire alarm, elevators, and computer networking would be appropriate. For the experienced electrician textbooks will supply needed information. If the field is of interest, you should be able to work into it in an incremental way.

> If a fluorescent fixture is out, it is usually the bulbs. If it still will not light, verify that there is power to the ballast. If so, disconnect power, cut the wires near the ballast, mount a new ballast, and reconnect using the cut wires as a guide. If you lose track, there is a wiring diagram printed on the ballast, as well as the number and type of bulbs for which the ballast is good. If you upgrade from T-12 to the more energy efficient T-8 bulbs, the existing sockets are compatible with either bulb, but the ballast must be changed. Running bulbs beyond the point where the ends are dark causes them to draw more current, and overheat the ballast. It is more economical to change bulbs on a regular schedule rather than waiting for them to burn out.

Take nothing for granted! At one time, I was called upon to add some fluorescent light fixtures, with new branch circuits and switch loops, in a small hospital located in a suburban area. Despite the fact that the building was less than 10 years old, I found the existing wiring to be highly substandard. Branch circuits were installed above a suspended ceiling in a general care area, where I found covers left off of junction boxes and receptacle circuits run in 12-2 type NM (Romex) cable tied to copper tubing that carried oxygen. Additionally, patient bed locations lacked the required number of receptacles and they were not identified.

The lesson to be derived from all of this, again, is: Take nothing for granted! If the existing installation is deficient in any way, you, your coworkers, the healthcare facility personnel, and the patients can be endangered. If you do any work that is appended to a substandard installation, unintended defects may surface even if your own work is impeccable. For this reason and others, it is essential that before and during any job you do in a healthcare facility and after the work is finished, you review, inspect, evaluate, and, as needed, correct the electrical environment.

Outside of that, and given that you have a thorough understanding of the electrical requirements for a healthcare facility, most troubleshooting and repair is straightforward. Circuits are identified and numbered, and there should be documentation on file for all electrical equipment, whether it is an exercise machine in the physical therapy unit or the control panel for the fire alarm system.

The priority should be preventive maintenance, rather than run to failure. Clamp-on ammeters and temperature probes are valuable instruments for checking the condition of motors, and large main lugs at service entrance panels—wherever there is heavy current. Exercise breakers, vacuum out transformers with power locked out, test emergency lights, change fluorescent bulbs at scheduled intervals before the ends darken (indicating that they are drawing excessive current, heating the ballasts), and speak to operating personnel about any concerns they may have. Do not forget the backup generators. For the prime mover, check fluids, filters, and belts and perform other maintenance outlined in the operator's manual.

While these preventive maintenance operations are being conducted, written documentation should be generated and kept in files. Posted at large motors and critical equipment should be maintenance logs, with space for date, time, pertinent comments, and worker signatures so that damaging trends can be spotted early and corrective measures taken before costly damage occurs or tragedy strikes.

Troubleshooting and repair are not much different from that in smaller and less complex occupancies. The usual operator interview, half-splitting diagnostic procedure, visual inspection, print and online research, and similar techniques should suffice. The only difference is that in a healthcare facility where lives are at stake, all work must be checked and rechecked to make sure that there is no hidden flaw. It is useful, in such a sensitive setting, to work with knowledgeable colleagues and to get second opinions at critical points.

Solar and Wind Power

Wind turbines (see Fig. 15-1) are actually solar powered in the sense that because the sun heats the Earth's surface unevenly, pressure differentials are established, making the wind blow. Both solar and wind power are intermittent. For stand-alone applications, energy must be stored, usually in batteries. Where there is utility power, the grid becomes the storage mechanism. When it is dark or becalmed, the utility supplies electricity in the normal fashion, and when the customer generates excess, it reverse-meters back into the energy pool, appearing as a credit on the utility bill.

A solar PV array produces dc, which has to be converted to ac to work with the utility. This ac has to be the same voltage and frequency as the grid power, and the waveforms have to be locked in synch, the peaks occurring simultaneously. Wind turbines at one time contained dc generators, but at present they are mostly equipped with alternators, which, without heavy current-carrying brushes, are less expensive to implement and maintain. The output is three-phase ac. It is known as "wild ac" because the frequency, being speed dependent, fluctuates due to changes in wind speed. This unregulated ac is not good for most applications, especially motors. However, a simple diode network will rectify it, producing pulsating dc, which can be smoothed out by means of a capacitor.

For stand-alone installations, the dc is usually connected to a battery bank for storage and to stabilize the voltage. An important part of the equation is a series-blocking diode, which prevents the batteries from discharging back through the wind turbine when there is no wind.

Stand-alone solar PV systems are configured in a similar manner, with batteries, but the output is dc and does not need to be rectified. The rationale for stand-alone systems, wind or solar, is that they are too far removed from utility power to justify the cost of building a line extension.

In a stand-alone system, loads can be dc or ac, or conceivably a combination of the two. If the voltages match, dc loads are connected to the battery bank, through a distribution panel with a main disconnect and overcurrent protection for each branch circuit. A full range of dc appliances is available, at somewhat greater cost than are their ac counterparts, not because they are more expensive to manufacture but because they are specialized products.

For those who already own ac appliances or who value the convenience of plugging in standard equipment, the dc can be converted to ac, 120/240 volts, 60 Hz (50 Hz in Europe and many countries). The conversion equipment is known as a "power inverter."

Figure 15-1   Wind turbine at a resort hotel.

The original power inverters were rotary machines, with a round housing and mounting hardware, resembling electrical motors but with no output shaft, just closed bell housings at both ends. A single shaft mounted on bearings inside had dc motor windings and ac generator windings, with an enclosure mounted on the housing for wiring connections, two-wire dc input and a three-wire 120/240-volt ac output. Due to the rotary nature of the generator section, the output was a pure ac sine wave. Speed was regulated so that a constant voltage and frequency were maintained. It was similar to three-phase conversion equipment, which is necessary because there is no way to get three-phase power out of a single-phase supply using any type of transformer configuration.

The rotary power inverter ran quietly and reasonably efficiently, but the electronic power inverter has supplanted it because it is less expensive. At one time, this equipment was capable of producing only a square-wave output, as opposed to a sine wave. The voltage changes between positive and negative were instantaneous, the highs and lows being plateaus and flat fields rather than rounded peaks and valleys. This waveform was not good for ordinary uses. Motors supplied by square-wave power overheat and quickly cut out if thermally protected, or have a greatly shortened life. Moreover, the fast rise and fall times make for powerful harmonics, interfering with sensitive electronic equipment and making for dangerous heating of neutral conductors.

Over the years, solid-state inverter technology improved, providing increasingly closer approximations of a true sine wave. At present, the output closely resembles utility power.

Joining the Grid

For non-stand-alone applications, where utility power is at the site or it is feasible to do a line extension, another configuration becomes possible. This is known as "cogeneration." The wind or solar power is connected to the utility line. Depending upon the wind or solar output at any given time and the amount of connected load, there will be power drawn from the utility or a surplus flowing back to the utility. The step-down transformer, reverse fed, becomes a step-up transformer and surplus power is sold back to the utility. Metering is capable of measuring the power flow in either direction, and the utility bill reflects the balance. This arrangement works well for the home or business owner, in large part because there is not the great ongoing expense of setting up and maintaining the battery bank. It is essential that automatic transfer switch capability be part of the picture, so that during an outage utility workers will not be exposed to dangerous voltages backfed through transformers.

For the interactive system to work, the two waveforms must be exactly synchronized, the negative and positive peaks occurring simultaneously. This is accomplished by means of a synchronous inverter, which samples the utility voltage at all times and changes the wind turbine or solar PV output as needed.

There are many types of grid-tie inverters, with vastly different power ranges and programmable features with accompanying user interfaces. The performance requirements are similar. They may have high-frequency transformers, 60-Hz transformers, or they may be transformerless. Synchronous inverters for wind and solar power play the same role in both systems, but they are not always interchangeable. Wind systems, with much faster speed and voltage fluctuations and abrupt peaks, not to mention the wild three-phase ac, require a specialized synchronous inverter. The wind power inverter may be used in a solar PV system but a solar inverter, even if it has adequate rating, will at best give poor performance due to its slower response to abrupt voltage changes.

A long, slender, straight blade screwdriver is sometimes known as an electrician's screwdriver, with good reason. It is essential for getting into tight places and provides a lot of control for tightening the cable connector in a deep box, and similar tasks. Since it has such a small shank, you can spin it between thumb and forefinger and quickly unscrew long-threaded bolts.

In recent years, the trend in wind and solar system marketing has been to provide complete packages, so that the synchronous inverter exactly matches the application. If, however, the system is being assembled or repaired using off-the-shelf or surplus components, it is important to correctly configure the project.

The synchronous inverter cannot deal with the wind turbine's wild three-phase ac, although some advanced models contain inverter and rectifier within the same enclosure. Some wind turbines have an internal rectifier. If neither of these is the case, a separate external rectifier must be installed. Moreover, the peak wind-turbine voltage must not exceed the synchronous inverter's nameplate capacity, and it is prudent to figure in a 20 percent safety margin. Some inverters contain diode protection to guard against overvoltage, but this should be regarded as a supplemental protection only because diode failure can take the form of open- or short-circuit events, ending in inverter damage.

An advanced synchronous inverter incorporates, as part of the user interface, an alphanumeric display. It can show any of a number of failure codes, and the user's manual will provide translations and additional information. Some of these error codes pertain to the inverter's inner circuitry while others are responses to system-wide anomalies such as irregular utility power.

The inverter should be tested at regular intervals to make sure that it is capable of isolating the system from outside utility lines when there is an outage. The requirement is that the inverter functions like a transfer switch to interrupt both legs of a single-phase service when utility power goes down. The protection is implicit because of the way the inverter matches the utility service, but one can conceive of a hardware fault in the inverter that would allow power to flow through, so extra vigilance is recommended.

A solar PV system is more benign in terms of power output, but there are other hazards that workers need to keep in mind at all times (see Fig. 15-2). Remember that solar panels are always live when not in darkness, from the time they are removed from their packaging and even after decommissioning. Some workers put opaque covering over solar arrays before working on the system upstream from the first disconnect. This protection technique is not very reliable, nor is the protection afforded by darkness, because distributed system capacitance and utility or battery backfeed can unexpectedly energize parts of the system.

FIGURE 15-2 Solar PV array.

The blocking diode could be shorted as well. The best procedure when working on a solar PV system is to avoid touching any ungrounded wiring terminals or conductive surfaces even if you believe they are not live. When working on a roof, expect the unexpected.

Similar hazards exist in a wind power system. It is understood that any tower work should be done when there is no wind. Even then, both rotor and yaw should be mechanically locked in case of an unexpected gust.

The other hazard is from electrical backfeed. Battery bank and utility lines must be effectively locked out to prevent shock hazard. A dc generator, in an older system, will "motor" when backfed, and a blocking diode is placed in series to prevent this from happening when there is no wind. A normally open pushbutton, labeled MOTOR, may be shunted across this diode for testing purposes. It is easy to imagine a situation where the diode failed or the button was pushed at the wrong time, causing a blade to strike a maintenance worker on the tower.

Troubleshooting of wind and solar PV systems is simple conceptually because of the way components are connected in discrete stages and because symptoms point directly to causes. However, the actual process is made difficult due to problems of access and accompanying dangers. These factors become more pronounced in larger systems, particularly in utility-scale wind farms and solar installations.

Starting at the small end, we will consider some maintenance and troubleshooting issues that arise in connection with a stand-alone wind power system. As in all such installations, the best way to minimize downtime, extended outages, and high repair costs is by instituting a preventive maintenance program. This is preferable to a frugal run-to-failure policy, which in the end will be far more expensive and can result in human injury or worse.

Stand-alone wind power systems are usually relatively small, although in remote areas where there is no possibility of a grid connection, some fairly large facilities could be served. Wind power is an attractive and viable solution for isolated research and industrial projects where the cost of moving in diesel fuel, as an alternative, would have to be factored in.

Batteries (When Needed)

In stand-alone systems, a big maintenance item is the battery bank. They are expensive, essential for good performance, and demand more maintenance than other parts of the system. Like solar panels, they are always on. When working around batteries you need to be aware of the dangers. A single cell is a low-voltage unit, but in series combination, there is a palpable shock hazard. In parallel, awesome arc fault energy becomes available, so there is no room for error. This is very true even for a small residential installation, and in a large commercial setting the hazard is multiplied.

Adequate space with good ventilation should be provided, so this has to be considered as part of the cost equation. Batteries lose their charge as time elapses. They are inherently inefficient, and the larger the battery bank, the greater the loss, so it should be sized just right, not too small and not too large. For this and other reasons, battery banks should be contemplated only for off-grid installations, not in conjunction with utility-interactive systems.

In wind and solar off-grid systems, lead-acid batteries are the way to go. Lithium-ion, NiCad, and other exotic types are not at all cost competitive.

Lead-acid batteries contain a liquid electrolyte that is a mixture of water and sulfuric acid. The greater the charge in the battery, the more acidic is the mixture. The amount of charge in a battery cannot be measured directly by a voltmeter. A hydrometer is used. It samples the electrolyte, measures its specific gravity, and translates that into percentage of charge.

While being charged, lead-acid batteries release hydrogen. If the rate of charge is fast and there are many batteries, the hydrogen will accumulate faster than it disperses, and there will be an explosion if there is a spark to ignite the gas. If there is a prolonged series of rapid charge–discharge cycles, a battery will overheat, possibly causing an internal explosion, rupturing the case, and releasing a blinding shower of strong acid.

The electrolyte level in lead-acid batteries can go down with use and when the tops of the plates are no longer submerged, there is damage. When lead-acid batteries are overheated, the plates warp and get too close to one another at some points, which shows up as an internal partial short and subsequent inability to hold a charge. When lead-acid batteries remain discharged for a period of time, the plates become corroded. This is called sulfation and manifests as an inability of the battery to take a charge. A sulfated battery can sometimes be partially restored by sharply rapping on the side of the case, causing the corroded material to fall off the plates and accumulate harmlessly at the bottom. In addition, a mild case of sulfation can sometimes be remedied by applying a long dc charging voltage.

At best, batteries deteriorate with age and will need to be replaced every few years, while a synchronous inverter will last much longer, making cogeneration where possible the better choice.

Battery posts and cables are subject to corrosion and should be taken apart and cleaned periodically, using a wire brush. I have used corrosion inhibitor to good effect, the type that is used for aluminum wire terminations in services. When reattaching clamps, torque the bolts a sensible amount. Before doing anything, remove both load and supply voltages so that there will not be arcing, in case there is hydrogen present. However, do not disconnect the supply when the wind is blowing without first inserting a dummy load, as the rotor speed and voltage will go high if there is no impedance to cut them down.

Lead-acid batteries last longer if they are given an equalization charge once a week. This is about 10 percent higher than the ordinary charging voltage, and you should apply it for 8 hours. The purpose is to intentionally create gas bubbles, which mix the electrolyte and prevent sustained layering, which makes a more acidic mix at the top, shortening battery life. Equalization charging is not necessary in an automobile because bumps in the road prevent layering in the battery.

In a stand-alone wind or solar installation, the battery bank requires more preventive maintenance than other parts of the system, but nevertheless there are many items that should be checked. For an older existing stand-alone wind turbine system that appears to be functioning efficiently, one would want to perform a careful overview in preparation for instituting a comprehensive preventive maintenance program. We should ascertain whether the system is NEC compliant. Assuming the backup battery bank has been checked and serviced as necessary, we will want to refer to NEC Article 694, Small Wind Electric Systems, to see if any critical requirements were overlooked in the original installation. The Code is revised every three years, and many new requirements are added and new areas of applicability are attached to existing requirements. The trend is toward more regulation, although in a few cases rules are relaxed somewhat or deleted entirely when they are judged not necessary from a safety point of view. Despite changes, it is not contemplated that there will be an immediate alteration of existing structures to conform to the new language. Electrical installations must comply with the edition of the Code that was in force in the jurisdiction at the time of construction, unless specifically stated to the contrary.

Nevertheless, the owner, in consultation with on-site or outside electricians and professional technicians, always has the option to upgrade the installation to comply with the latest standards. Moreover, Code rules are minimum requirements and, in many cases, it is feasible to go beyond them in the interest of increased safety and efficiency and to allow for future expansion. This is especially true for wind and solar projects. An example is an underground power line that runs from a wind turbine tower to the building. It may be judged desirable to install an oversized conduit (PVC is generally used) to accommodate larger conductors that may be needed in the future, or indeed to put in the larger line in the first place. These upgrade options, the second much more expensive than the first, will never be regretted once the trench is backfilled, graded, and reseeded.

Another area where more is better is grounding. Here it is possible to make significant upgrades at modest cost. In solar and even more so in wind power installations, damage from lightning is a distinct threat, especially in areas of high ground resistance. Many wind turbine installers just throw in an extra ground rod and call it good, not realizing that an undersized and floating ground electrode of this sort may do more harm than good. The central concept is bonding, and this goes way beyond ground rods. The Code provides information on this and other aspects of electrical safety regarding solar PV and wind turbine installations. We will look at wind power requirements first because they are more complex, and then see how they differ in the context of solar electrical power.

It is noted in Part I, General, that Article 694 applies to small wind electric systems that consist of one or more wind electric generators having a rated power up to and including 100 kW. This would include a wind farm situation where the installation powers a single home or facility and everything is under single ownership. If the power production and distribution equipment is privately or publicly owned and power is supplied to customers, it is not under NEC jurisdiction, but is regulated by a public utilities commission, and design and installation details are covered by the National Electrical Safety Code.

A stand-alone or grid-tie system may consist of any number of wind turbines installed at one location. The usual setup, however, is a single unit, and the size is often well below the 100-kW limit. Therefore, it is under NEC jurisdiction.

Section 694, Definitions, gives the meanings of terms specific to this article.

- **Charge controller** is equipment that controls dc voltage or current or both used to charge batteries. Including a good charge controller, where there are batteries, is essential for a safe and reliable installation because overcharging will damage the batteries and release quantities of hydrogen, setting the stage for fire or explosion.

- **Diversion charge controller** collaborates with the charge controller to protect the battery bank from overcharging. It diverts excess electrical energy to a dummy or other load when the batteries are fully charged and production exceeds demand. If possible, this energy should be put to good use, such as heating a building in cold weather, but if that is not possible, it will be necessary to dump or dispose of the excess energy to prevent damage to the batteries.

- **Diversion load** is the dump load used in connection with the process described above.

- **Guy** is a cable that mechanically supports a wind turbine tower. Some towers do not require guy wires such as when the foundation is of sufficient mass and depth to support the tower against the side thrust of the most powerful wind that will be encountered and when the tower itself is strong enough so that it will not buckle in heavy winds.

- **Inverter output circuit** is made up of the conductors between an inverter and an ac panelboard for stand-alone systems, or the conductors between an inverter and service equipment or another electric power production source, such as a utility, for an electrical production and distribution network.

- **Maximum output power** is defined as the maximum 1-minute average power output a wind turbine produces in normal steady-state operation. Instantaneous power output can be higher. This is an all-important parameter because it determines sizing of the inverter, rectifier, and associated wiring.

- **Maximum voltage** is defined as the highest voltage the wind turbine produces in operation including open-circuit conditions. This figure also influences the choice of inverter and rectifier. Normally, the wiring insulation is rated for 600 volts.

- **Nacelle** is an enclosure housing the alternator and other parts of the wind turbine. Among considerations are that it have aerodynamic form, provide protection for its contents against often extremely hostile elements, and have a good appearance from an aesthetic point of view.

- **Rated power** is defined as the wind turbine's output power at a wind speed of 24.6 mph. If a wind turbine produces more power at lower wind speeds, the rated power is the wind turbine's output power at a wind speed less than 24.6 mph that produces the greatest output power.

- **Wind turbine output circuit** is defined as the circuit conductors between the internal components of a small wind turbine (which might include an alternator, integrated rectifier, controller, or inverter) and other equipment.

Section 694.7, Installation, makes some points that pertain to this subject. It is stated that wind turbine systems are to be installed only by qualified persons. Exact mechanism for licensing or certifying is not specified. Elsewhere (Article 100, Definitions) in the Code, *qualified person* is defined as one who has skills and knowledge related to the construction and operation of the electrical equipment and installation and has received safety training to recognize and avoid the hazards involved.

This section further provides that a small wind electric system employing a diversion load controller as the primary means of regulating the speed of a wind turbine rotor is to be equipped with an additional, independent, reliable means to prevent overspeed operation. An interconnected utility service is not considered a reliable diversion load.

Another installation requirement for small wind turbine systems is that a surge protective device is to be placed between the installation and any loads served by the premises' electrical system. Types that are permitted are enumerated.

Receptacles are permitted to be supplied by a small wind electric system branch or feeder circuit for maintenance or data acquisition. They are to be protected by an overcurrent device with a rating not to exceed the current rating of the receptacle.

Part II, Circuit Requirements, contains sections on maximum voltage and overcurrent protection. For wind turbines connected to one- and two-family dwellings, turbine output circuits are permitted to have an output voltage up to 600 volts. Higher voltages are permitted for other occupancies. The installations must comply with Part IX of this article, titled Systems over 600 Volts. That part states that in battery circuits, the voltage used is to be the highest voltage experienced under charging or equalizing conditions.

In addition, it is further provided that small wind turbine electric system currents be considered continuous. This is logical because frequently the wind will blow without interruption for more than 3 hours.

It is stated that small wind turbine systems are to have overcurrent protection in accordance with Article 240, which lays out the principles regarding overcurrent protection for conductors and electrical equipment in general. (Some exceptions are noted later. Fire pumps, for example, in the context of motor overload are to be run to failure rather than shut down during firefighting operations.)

The inverter output of a stand-alone small wind electric system is permitted to supply 120 volts to single-phase, three-wire, 120/240-volt service equipment or distribution panels where there are no 240-volt outlets and no multi-wire branch circuits. This is a common arrangement. The one hot, ungrounded wire from the inverter is connected to both busbars in the distribution panel. Consequently, hooking up a multi-wire branch circuit would result in an unbalanced neutral, which would overheat and create a fire hazard.

Mandatory Disconnects

Part III, Disconnecting Means, should be read carefully and the small wind turbine system should be checked for full compliance before performing any troubleshooting or repair. Remember that the leads coming out of the turbine must be regarded as live notwithstanding the blocking diode, even when the rotor is not turning. The disconnect protects maintenance workers when open by isolating live portions of the circuitry.

The first and foremost rule is that means must be provided to disconnect all current-carrying conductors of a small wind electric power source from all other conductors in a building or other structure. A switch, circuit breaker, or other device, either ac or dc, is not to be installed in a grounded conductor if its operation leaves the grounded conductor in an ungrounded and energized state. An exception provides that a wind turbine that uses the turbine output circuit for regulating turbine speed does not require a turbine output circuit disconnecting means.

Several other requirements are listed for small wind electric systems regarding the disconnect capability. The disconnecting means does not have to be suitable as service equipment. However, it is to consist of manually operable switches or circuit breakers that comply with all of the following requirements:

- They must be readily accessible.
- They must be manually operable without exposing the operator to live parts.
- They must plainly indicate whether they are in the open or closed position.
- They must have an interrupting rating sufficient for the nominal circuit voltage and the current that is available at the terminals of the equipment.
- Where all terminals of the disconnecting means are capable of being energized in the open position, a warning sign must so state.

It is noted that equipment such as rectifiers, controllers, and output circuit isolating and shorting switches are permitted on the wind turbine side of the disconnecting means. There are several requirements. The small wind electric system disconnecting means is to be installed at a readily accessible location either on or adjacent to the turbine tower, on the outside of the building or structure served, or inside at the point of entrance of the wind system conductors. The wind turbine disconnecting means is not required to be located at the nacelle or tower. It is not to be located in a bathroom.

As in all electrical disconnect situations, small wind system disconnects are to consist of not more than six switches or circuit breakers mounted in a single enclosure, in a group of separate enclosures, or in or on a switchboard.

Rectifiers, controllers, and inverters are permitted to be located in nacelles or other exterior areas that are not readily accessible.

Means are to be provided to disconnect equipment, such as inverters, batteries, and charge controllers from all ungrounded conductors of all sources. If the equipment is energized from more than one source, the disconnecting means are to be grouped and identified.

A single disconnect is permitted for the combined ac output of one or more inverters in an interactive system. Equipment housed in a turbine nacelle is not required to have a disconnecting means.

Part IV, Wiring Methods, specifies that all turbine output circuits in readily accessible locations that operate at over 30 volts be installed in raceways.

Flexible cords and cables, where used to connect the moving parts of turbines or where used for ready removal for maintenance and repair, are to be of a type identified as hard service or portable power cable, suitable for extra-hard usage, listed for outdoor use, and water resistant.

Direct-current turbine output circuits installed inside a building or structure are to be enclosed in metal raceways or installed in metal enclosures from the point of penetration of the surface of the building or structure to the first readily accessible disconnecting means.

To summarize, in a small wind electrical system, everything has to be capable of disconnection unless it is located in the nacelle and everything up to the main disconnect inside has to be in metal raceway unless it is not over 30 volts or it is not accessible.

Part V, Grounding, provides that exposed noncurrent-carrying metal parts of towers, turbine nacelles, other equipment and conductor enclosures are to be connected to an equipment grounding conductor regardless of voltage. Attached metal parts, such as turbine blades and tails that have no source of electrical energization are not required to be connected to an equipment grounding conductor.

A wind tower is to be connected to one or more auxiliary grounding electrodes. This is in addition to the full-scale bonding by means of the equipment-grounding conductor that is required to be made to the overall electrical grounding system.

Electrodes that are part of the concrete foundation are acceptable and in fact make an excellent grounding electrode system. The rebar must not have an insulating coating and the concrete must be in direct contact with the earth, not resting on plastic or foam. Contact of dissimilar metals is to be avoided anywhere in the system to eliminate the possibility of galvanic action and corrosion.

Auxiliary electrodes and grounding electrode conductors are permitted to act as lightning protection system components. If separate, the tower lightning protection system grounding electrodes are to be bonded to the tower auxiliary grounding electrode system.

Guy wires used as lightning protection system grounding electrodes are not required to be bonded to the tower auxiliary grounding electrode system.

These are the principle NEC requirements for a small wind electric system. A wind system is substantially more complex than a solar PV system due to its demanding mechanical components and rapidly fluctuating output voltage. The solar PV system does not have moving parts unless there is a tracking system and since there is not a tall tower, lightning and violent storms are less of a concern. A more stable dc output, as opposed to the wind turbine's wild ac output, permits a simpler electrical system. That being said, we must recognize some hazards unique to solar PV systems.

Like Article 694, Article 690, Solar PV Systems, begins with definitions. The principle terms, excluding those covered in Article 694, are as follows:

- **Alternating current module** is a complete, environmentally protected unit consisting of solar cells, optics, inverter, and other components exclusive of tracker, designed to produce ac power when exposed to sunlight.

- **Array** is defined as a mechanically integrated assembly of modules or panels with a support structure and foundation, tracker, and other components, as required, to form a dc power-producing unit.

- **Bipolar photovoltaic array** is a photovoltaic array that has two outputs, each having opposite polarity to a common reference point or center tap.

- **Blocking diode** is defined as a diode used to block reverse flow of current into a photovoltaic source circuit.

- **Building integrated photovoltaics** is defined as photovoltaic cells, devices, modules, or modular materials that are integrated into the outer surface or structure of a building and serve as the outer protective surface of that building.

- **Module** is a complete, environmentally protected unit consisting of solar cells, optics, and other components, exclusive of tracker, designed to generate dc power when exposed to sunlight.

- **Monopole subarray** is defined as a PV subarray that has two conductors in the output circuit—one positive and one negative. Two monopole PV subarrays are used to form a bipolar PV array.

- **Panel** is a collection of modules mechanically fastened together, wired, and designed to provide a field-installable unit.

- **Photovoltaic output circuit** is defined as circuit conductors between the photovoltaic source circuits and the inverter or dc utilization equipment.

- **Photovoltaic power source** is an array or aggregate of arrays that generates dc power at system voltage and current.

- **Photovoltaic source circuit** is defined as the circuits between modules and from modules to the common connection points of the dc system.

- **Photovoltaic system voltage** is the dc voltage of any photovoltaic source or photovoltaic output circuit.

- **Solar cell** is defined as the basic photovoltaic device that generates electricity when exposed to light.

- **Solar photovoltaic system** is the total components and subsystems that, in combination, convert solar energy into electric energy suitable for connection to a utilization load.
- **Subarray** is an electrical subset of a PV array.

The above NEC definitions demonstrate that the structures of wind and PV systems are quite similar. Both are used in stand-alone or utility-interactive configurations. There are some differences. Solar power requires no rectification because pure dc is what is produced. However, solar systems do require extensive parallel and series connections because the voltage and current output of a single cell is small. Both solar and wind installations usually have battery backup for stand-alone applications and they have synchronous inverters when cogenerating with a utility.

Part II, Circuit Requirements, in a section on maximum photovoltaic system voltage, makes the point that voltage for a given system has to be corrected for ambient temperature. We ordinarily think of chemical and physical processes as becoming more active as the temperature increases. Solar PV cells actually have greater electrical output at lower ambient temperatures. This effect is most pronounced for the open-circuit voltage. An increase in temperature reduces the band gap of the PV semiconductor cells, so at higher temperature there is less output. Because open-circuit voltage is the parameter of interest in determining the voltage rating of cables, disconnects, overcurrent devices, and other equipment, the Code provides a table, 690.7, that gives voltage correction factors for crystalline and monocrystalline silicon modules. These correction factors are for ambient temperatures below 77°F, ranging down to –40°F. The procedure is to decide on the lowest expected ambient temperature for your region, then multiply the open-circuit voltage by the correction factor from the table for that ambient temperature. The resulting corrected open-circuit voltage is the value used to specify the system components mentioned previously. As examples, the correction factor for an ambient temperature of 50°F is 1.06, while the correction factor for an ambient temperature of –20°F is 1.21, so we see the temperature effect is quite pronounced.

Section 690.8, Circuit Sizing and Current, states that the maximum current for a specific circuit is to be calculated as follows:

- The maximum current for photovoltaic source circuits and photovoltaic output circuits is the sum of parallel module rated short-circuit currents multiplied by 125 percent.
- The maximum current for inverter output circuits is the inverter continuous output current rating.
- The maximum current for a stand-alone inverter input circuit is the stand-alone continuous inverter input current rating when the inverter is producing rated power at the lowest input voltage.

Section 690.9, Overcurrent Protection, provides that photovoltaic source current, photovoltaic output circuit, inverter output circuit, and storage battery conductors and equipment are to be protected in accordance with the requirements of Article 240, Overcurrent Protection.

Section 690.11, Arc-Fault Circuit Protection (Direct Current), provides that photovoltaic systems with dc source circuits, dc output circuits, or both, on or penetrating a building operating at a PV system maximum system voltage of 80 volts or greater are to be protected by a listed arc-fault circuit interrupter, PV type.

Part III, Disconnecting Means, requires disconnecting capability for all parts of a PV system, and the requirements parallel those for small wind electric systems.

Part IV, Wiring Methods, is similar to the corresponding part for wind power systems, notably in the requirement that where photovoltaic source and output circuits operating at maximum system voltages greater than 30 volts are installed in readily accessible locations, circuit conductors are to be installed in raceway. Since PV raceways cannot be directly attached to PV modules, complying with this requirement could present a problem. In most cases, however, the modules are in a location that is not readily accessible such as a rooftop, where a ladder is required for access. If the modules are readily accessible, it may be necessary to fence the area with a locked gate to comply with this rule.

Wiring methods for solar PV systems in the areas that are not readily accessible are unique in that single-conductor cable, generally not Code compliant, is permitted in this application. Single-conductor cable Type USE-2 and single-conductor cable listed and labeled as photovoltaic wire are permitted in exposed outdoor locations in photovoltaic source circuits for photovoltaic module interconnections within the photovoltaic array.

Negative and positive conductors of each circuit as well as the equipment grounding conductors should, however, be run as close as possible to minimize the time constant of the circuit and to minimize induced currents from lightning that may strike in the area.

Generally, the Code prohibits the use of conductors smaller than 14 AWG, but for module interconnection where ampacity requirements are met, 16 AWG and 18 AWG wire may be used.

Wiring methods are not to be installed within 10 in. of roof decking or sheathing except where directly below PV modules and associated equipment. Circuits must be run perpendicular to the roof penetration. The requirement is to prevent damage from saws used by firefighters to vent a roof. Remember that these conductors are always live and not de-energized by the disconnect.

Part V, Grounding, provides that for a photovoltaic power source, one conductor of a two-wire circuit with a photovoltaic system voltage over 50 volts and the reference (center tap) conductor of a bipolar system are to be solidly grounded. The dc circuit grounding connection is to be made at any single point on the photovoltaic output circuit. Locating the grounding point as close as practicable to the photovoltaic source better protects the system from voltage surges due to lightning.

The battery bank requirements for a stand-alone system parallel the requirements for small wind electric systems.

The purpose for including a summary of the NEC coverage of wind and solar PV systems is to remind the technician that before commencing an extensive troubleshooting or repair procedure, Code compliance should be verified. The importance of this aspect of the installation cannot be overstated because this is how we prevent hazards.

Troubleshooting and repair procedures are not overly difficult in wind and solar installations once the hazards are recognized and proactive measures taken to deal with them. To emphasize, do not make any assumptions regarding potentially live parts. In addition, for a wind system, do not climb a tower unless the weather is calm (including no possibility of lightning) and rotor and yaw are locked mechanically so they cannot move.

With these cautions always before us, we will move on to some maintenance and troubleshooting procedures, beginning with wind systems. Assuming the battery bank has been inspected and serviced as outlined earlier, we may move on to other parts of the system. A thorough maintenance schedule should be instituted, with everything documented so that items are not omitted.

Aside from the battery bank, the maintenance interval is generally taken to be six months. This is in addition to ordinary walk-around visual surveillance and observation of any electrical meters, pilot lights, and the alphanumeric display that is part of the inverter's user interface. Additionally, the operator should be conscious of any unusual noises that might indicate there will be trouble down the road. Moreover, if there is good wind but there seems to be a falling off of electrical power, an investigation is in order. We will discuss this and other specific symptoms along with possible causes.

The first priority for a preventive maintenance program is the tower. If there is any weakness, it will only get worse and disaster lies on the horizon. The heaviest wind is the one that counts, and it is going to exert enormous side pressure at the top of the tower. Even if there are only two blades, the swept area should be visualized as a solid circular plywood disc, where the entire force of the highest wind that strikes it wants to push over the tower because that is the effect when the rotor is spinning at high speed.

The tower should be examined carefully, starting from the ground. Check the concrete foundation for any cracking or surface deterioration. Make sure that erosion or settling has not reduced the amount of fill around the perimeter of the concrete, and see if the top surface is level or pitched slightly to avoid standing water. If there are anchor bolts, check that they are tight and not corroded.

If there are guy wires, make sure that there is reasonable tension, equal on each, and that there are no nicks or cuts that would make a weak point. The guy wires should be firmly anchored, with no signs of rust.

Work your way up the tower, checking for any deterioration of the metal, bending, loose bolts, etc. Bolts should be torqued to specifications that come with the tower if it is a manufactured unit. Otherwise, generic torque specifications are available in a mechanical engineering textbook or online. If the bolts are under-torqued, they will loosen with wind and turbine vibration. If they are over-torqued, they will break more readily or the threads will become worn, setting the stage for loosening.

Wind turbines vary greatly with make and model. Specific maintenance guidelines will be found in the owner's manual. If it is missing, try to find a copy online or contact the manufacturer.

The blades encounter a lot of stress, and should be inspected carefully. If they are wood, they must be repainted periodically. In addition, a leading-edge tape is usually applied. Its purpose is to protect the blades from damage due to abrasive material and insects in the air. It may be possible to repair damage such as small cracks and surface damage to the wood, but do not expect blades to last the life of the turbine. Where they attach to the hub, look for looseness and enlarged bolt holes.

The turbine has three separate moving mechanisms. The most dramatic of these is the spinning rotor. Second, in a horizontal axis wind turbine, there is yaw rotation, which is necessary to keep the machine facing into the wind as it changes direction. Finally, there is the unfurling mechanism that shuts down the turbine during excessively powerful winds when continued operation would invite damage. This action can consist of a spring-loaded spoiler that deploys due to centrifugal force when turning at high speed, using air resistance to cut the rpm, or alternately a tip-down mechanism so that the blades are no longer facing into the wind.

Each of these actions involves moving parts, requiring lubrication or other maintenance. The rotor generally has front and rear bearings that are part of the generator or alternator. They will require lubrication, or they will be sealed bearings. Many mechanics feel that grease fittings provide a more continuous supply of lubrication. The problem with grease fittings, however, is that they permit abrasive material to get into the bearings. The smallest grain of sand will scratch the metal surfaces, generating metal particles that make for even more wear. The answer is to be scrupulous in cleaning the fitting before applying the grease gun. If there is any accumulation of old grease and dirt around the fitting, scrape it away with a flat-blade screwdriver. Then wipe the fitting with a clean rag. Clean any excess grease off the end of the grease gun nozzle in case it has picked up some dirt.

The other problem with the grease fitting is over-greasing. Too much grease will break the bearing seals, allowing entry of dirt and moisture. Another problem with too much grease is that it makes for overheating and short bearing life. This is because the grease is under pressure and there is more friction. The answer to this is to pump in only a moderate amount of grease. If possible, it would be worthwhile to have temperature sensors (with remote digital readouts) on each bearing so that trouble can be spotted early.

The yaw mechanism will have a heavy low-speed thrust bearing that should be inspected for play, rough operation, and noise. When the bearing becomes worn, it should be changed before there is a disaster.

At one time, most wind generators were dc units. They produced a fluctuating dc that could be regulated and used for charging batteries without the need for rectification. However, the dc generators required large current-handling brushes, which meant higher initial cost and more maintenance, not only for brushes, but also for the commutator. Overall, the alternator, wild three-phase output notwithstanding, proved less expensive and more reliable.

When Output Declines

It is not possible to have a troubleshooting procedure that will cover all makes and models because there are so many variations. The best approach is to consult the manufacturer's documentation. However, here are a few fundamental generalizations. Throughout, we are assuming that basic battery maintenance is up to date, and all battery cable connections have been checked and cleaned where necessary.

If the power output is abnormally low, you have to observe the blades to see if they are turning at the correct speed. If they were turning too slowly, that would account for the reduced output. Start with the blades. If they are worn, the altered shape will result in poor performance, as would an accumulation of ice. If the blades are judged to be in good shape, the next thing to check is the bearings. If one of them is running hot, that is a sure indication that energy is being lost at that point. When there is no wind, turn the blades slowly and see if there is a gravelly feeling or roughness perceived. If so, is it continuous or does it appear at a single point in the rotation? If the latter, it is likely that something has come loose in the alternator, making an obstruction. A stethoscope is useful in locating the source of the noise. You can improvise, using a length of neoprene tubing or even a long screwdriver. Press the end of the handle against your ear and touch the blade tip to bearing and equipment housings.

If the blades are good and the turbine's drive shaft and alternator are turning freely, you have to look elsewhere. Check the yaw. If it does not swivel freely, it is possible that the turbine is not consistently facing into the wind.

If these investigations reveal that the problem is not mechanical, it makes sense to look for an electrical solution. If there is reduced electrical output, the troubleshooting process is straightforward and simply involves taking voltage and current measurements at various points and checking components. For example, you should measure a very low voltage drop across the slip rings. You will obviously feel uncomfortable taking meter readings at this location with the rotor turning. Consider attaching a piece of Cat 5e to the test points, and running it to the ground so that a voltage measurement can be made after the wind comes up.

Keep in mind that for a given wind speed and reduced or no electrical output, an open or high-impedance series fault between the alternator and battery bank or synchronous inverter will result in full rotation speed of the rotor, whereas a short, partial shunt fault or abnormally heavy loading will result in drag on the rotor, showing up as reduced rotor speed.

Naturally, you will want to begin at ground level. Unhook both ends of the underground line and do ohm readings with the far ends alternately not connected and shunted together. Line faults are usually caused by lightning, which can damage the underground portion. If the line is found not to have a fault, look at the components. If the blocking diode has become shorted, the batteries will discharge through the alternator, and if it is open, the batteries will not receive a charge.

Rectifying diodes can be inside the alternator or located outside. Either way, they will be fairly large and have substantial heat sinks. The connection between a diode and its heat sink resembles an electrical connection in that any looseness or exposure to corrosive agents will allow an oxide layer to form, and there will be poor heat transfer. The diode will overheat and quickly no longer be a diode. An off-the-shelf replacement will work if the forward and reverse bias voltages match and there is sufficient power-handling ability, measured in watts. If there has been a problem with diode failure in the past, it may be due to undersizing, especially of the heat sink. A diode should be located in a well-ventilated area to facilitate heat dissipation and never near the batteries where the corrosive atmosphere could cause heat sink failure. Discarded battery chargers and dc welders are excellent sources of replacement diodes with huge heat sinks. However, be aware that large diodes may have increased amounts of reverse bias leakage, which over the long haul could be costly due to battery discharge through the generator windings.

If it is judged that the alternator or generator has poor output or no output at all, it will have to be removed from the tower and brought into the shop, as a complete teardown will probably be on the agenda. Before disassembly, put alignment marks on the parts of the housing that are to be separated so they will go back together correctly. Two light parallel scores with a hacksaw blade work well and then there will be no uncertainty.

Disassemble the machine and check for any compromised wiring or loose components. You can do ohmmeter tests on the windings to check for open circuits, but do not expect this to be definitive. In no case should either end of any winding be grounded out. Resistance should be in the high megohm range.

Check for loose or noisy bearings, and clean any corrosion or dirt from all parts. A complete overhaul, involving rewinding, will require the services of a motor-generator rebuilding shop, using specialized materials and techniques. However, you can do many less exacting repairs, and often this will suffice to put the machine back in service.

For an old dc generator, begin by seeing if the shaft will turn freely. If you give it a good spin by hand, it should rotate several turns before coasting to a stop.

Many of these units have ceramic magnets fastened with epoxy to the inside of the generator housing. One or more of them may have become loose, causing noise and reduced output. Any loose magnet can be remounted using high-temperature epoxy. Be sure to duplicate the exact original position and orientation. Be careful using tools on these magnets as they are very brittle.

Brushes are a high-maintenance item. If there has been moisture infiltration or over-oiling of a bearing, the brush material can become softened causing rapid wear. Springs that hold the brushes firmly against the commutator may lose tension. Sometimes it is possible to stretch the springs by hand and restore operation. If one or more of the brushes has become too short, it should be replaced before there is damage to the commutator. Replacement brushes are available from the turbine manufacturer or as generic products. Some technicians have had success cutting down larger brushes to make something that fits. However, the exact material composition of the brush is also critical, so such improvisation can result in short brush life, damage to the commutator, and poor performance.

Check the commutator. If it is out of round (as measured by a micrometer) or disfigured in any way, the usual procedure is to take it to a machine shop and have it turned, removing as little material as possible. Afterward, the grooves will need to be undercut, using a tool made for the purpose that resembles a hacksaw blade with a handle.

Procedures for rebuilding an alternator are similar. A new alternator is enormously expensive, but it is always rebuildable. An alternator is a little more compact than a generator, so more care is needed. As always, begin by making alignment marks on the housing parts so they will go back together correctly.

Remove the through bolts, brushes and holders, and rear bearing. Clean the rear-bearing cavity. If there is an internal regulator, replace it if necessary. Check that the ground connections are not corroded. If installing a new bearing, steps have to be taken so that the race does not spin in the case. This can be done by using a punch to make small indentations in the race at three equally spaced points. Another method is to epoxy the race in place. Use only a small amount so that the space taken up by the epoxy does not prevent the case from going back together.

Check the stator windings. If the insulating paint is worn away, the windings might have to be replaced. The wire turns are very tight and rewinding is not feasible for a small unit.

After reassembly, spin the alternator by hand. It should turn freely, gradually coasting to a stop. There should be no scraping sound or roughness.

On the bench, the alternator can be rotated using an electric drill. Be sure to connect batteries or some other load. Check with a voltmeter for proper charging voltage.

If the wind system is utility interactive, the downstream end should be checked out. Generally, the rectifier and synchronous inverter are very reliable and will not require rebuilding unless there is lightning damage. Monthly utility statements, as a start, should be closely monitored to see if there is an unexpected drop in performance. A large synchronous inverter will have a user interface that includes an alphanumeric display. It will show fault codes that can be interpreted using the manufacturer's documentation, found in the user's manual or online.

Troubleshooting in this area is straightforward. Make sure all terminations and cables are in good order, and then go inside the enclosure to find out what is going on. You will need the schematic.

Rectifier and synchronous inverter outputs can be examined by means of an oscilloscope if there is a persistent problem.

Thus far, we have been concerned with wind and solar PV systems that occupy the small end of the scale. The greatest growth in the last few years, however, has been in utility-scale wind energy production. These installations are provocatively termed "wind farms." There is the immediate potential to meet 20 percent of our energy needs from this source alone, and eventually a much higher amount if large-scale energy storage becomes feasible.

It is clear, when we drive through rural areas, that these huge constructions are becoming more numerous, perched along mountain ridges where the wind is best. However, they are not without controversy. Opponents point to the fact that they intrude on pristine wilderness landscapes and, for those living close to a gigantic new installation, the constant low-frequency thumping sound that the turning blades make is bothersome to put it mildly. Moreover, because of their very nature, wind towers on the utility scale have to be located away from areas where the generated electricity is to be used. The good winds are not to be found in low-lying areas with tall buildings, nor is the land available. Remotely situated, wind farms require power lines, often with big steel towers. The answer to this is that all new transmission lines should be placed underground, and the property never taken by eminent domain.

I have to admit my bias. I like the looks of wind towers. I believe they complement the landscape, and I find their sound soothing. That being said, the rights of others to live in peace must be respected. Where there is opposition to the placement of a large wind project within the community, it should not be built. Perhaps large-scale solar PV, with low-lying, flat, very quiet arrays, is the better solution.

Nevertheless, as can be plainly seen, utility-scale wind farms are part of our life and will continue to be built. In view of that fact, we can expect to see an increased demand for maintenance and repair. As existing installations age and new ones become more numerous, there will be an increase in the number of workers required to keep these monstrous machines running.

The physical principles are the same for large-scale wind generation, but the actual machines are different. So are the people who work on them, and the firms that offer up-tower repair services. Workers fly all over the world on very short notice to replace bearings or do other large wind turbine repairs, laboring through the dark night and around the clock to get a disabled machine back online. Sometimes material and labor costs for these unscheduled repairs will exceed $100,000, and the only successful operating mode will involve a "whatever it takes" attitude that we find admirable.

The individuals who perform this work are different from many of us. Very courageous, they are highly paid mountain-climbing electrical engineers who have a Delta Force drive to get the job done.

You may never do this type of work, but if you want to broaden your perspective and gain some insight into this interesting world, read on. (It is best to explore areas outside our immediate domain on the theory that this will provide knowledge in situations we may encounter on the ground).

An interesting overview is provided by taking a brief survey of some of the huge utility-scale wind turbines that are currently available, with a glimpse at their inner workings. Then we will look at diagnostic programs that are available.

Big Machines

The late nineteenth century was a time of great advances in public usage of electricity, and it was in this period that the first steps toward large-scale production of electricity by means of wind occurred. In 1888 in Cleveland, Ohio, Charles Brush built a 56-ft multiple-blade wind turbine atop a massive tower. It had a large tail to provide yaw rotation. A dc generator turned at up to 500 rpm, thanks to a 50:1 step-up gearbox. With an output of only 12 kW, it remained online for 20 years and constituted the first tentative step toward utility-scale electrical production by means of wind power.

In 1891, Denmark was the site of a more efficient machine. Built by Poul La Cour, it had an advanced rotor with aerodynamic blades that turned at higher speed. Soon there were 25-kW machines in Europe, but falling oil prices and an unsettled economic environment discouraged further development, which came much later, in the 1920s in North America. In the Great Plains, farmers wanted electric tools and appliances, but many were still off-grid.

Jacobs and Parris-Dunn filled the need. Their designs incorporated advanced blades that resembled airplane wings rather than flat paddles. The surfaces developed lift as opposed to merely being pushed by the moving air, and the result was a more efficient means of capturing available wind energy.

Unfortunately for advocates of wind power, rural electrification spelled the end of an era. However, outside of North America, new developments surfaced in rapid succession. In Russia, in 1931, a 100-kW machine was built. This unprecedented output was still not large enough to justify such a massive piece of equipment, and the project was discontinued after two years. Soon thereafter, a much larger prototype was built in Vermont. A 1.25-mW machine was put up on Grandpa's Knob, and it operated successfully for a short time until a gust of wind tore off one blade, precipitating an out-of-balance disaster.

Several new designs with lightweight fiberglass and plastic blades appeared in the 1960s. Due to low oil prices, there was not much progress until a decade later, when rising fuel prices once again gave these technologies an economic advantage. A huge, 3.2 mW, 330-ft wind turbine was built in Oahu, Hawaii, and it was hoped that with the construction of many such units, it might be possible to lessen our dependence on fossil fuels.

Then, in the late 1980s and 1990s, first in Europe and eventually throughout the world, high quality, efficient, reliable wind machines for utility-scale use began to be manufactured. At present, the trend seems to be accelerating.

It is interesting to note that the installed cost of a single wind turbine is approximately $2 million per mW, and of course, full-scale wind farms have many such units. That is a lot of money, but then they make a lot of electricity in the life of the project. At retail costs, a medium-sized single wind turbine can produce over $500 worth of electrical power per hour when the wind is blowing.

To get an overview of how large wind turbines work, we will look at some typical makes and models currently available. Throughout this discussion, remember that the power produced is proportional to the square of the blade span and the cube of the wind speed. Accordingly, in the interest of electrical yield, manufacturers want to build ever-bigger machines and raise them higher off the ground.

The largest wind turbine maker in the world is Vestas. Its V-90 is a 3 mW giant with three blades and a blade span of almost 300 ft. The operating speed is 9 to 19 rpm. Pitch and speed control provide power regulation. During high-speed winds that could inflict damage, three air-operated cylinders provide braking action. Like most big machines, rpm is increased by means of a gearbox with one helical and two planetary stages.

General Electric is a major presence in the world of wind turbines. The 1.5 mW model has active pitch and yaw control. There is also power torque control with an asynchronous generator. A wind direction sensor controls yaw by means of a servo motor. Such high voltage and current precludes a slip-ring arrangement as in small electric wind machines. Instead, the main power cable, which also has various data conductors run with it, is simply allowed to twist. A turns counter keeps track of clockwise and counterclockwise rotation, and it directs the servo motor to rotate once in a while as needed to prevent an excessive twist. This model also has variable speed operation, so that the rotor can tolerate the highest wind speeds rather than shutting down and losing that energy.

Like all wind farm machines, the GE 1.5 mW model interacts with the grid. If grid power is not sufficiently stiff, there can be a problem. This machine features reactive power, with current leading voltage to stabilize voltage and increase transmission efficiency.

Another advanced capability is a reliable braking system, each blade having electromechanical pitch control and a hydraulic parking brake. Blade tips have lightning electrodes, minimizing this threat that is always present for wind systems. Active damping makes for reduced maintenance and increased life expectancy. There is less metal fatigue in the tower. Gearbox and generator have elastomeric support, which means quieter operation.

Siemens is another big player in the international large wind turbine-manufacturing arena. It is not the largest, but the SWT-2.3-93 2.3 mW wind turbine has an unusually large rotor, permitting it to capture energy at low wind speeds. There are no slip rings. It has automatic lubrication in addition to innovative safety features.

Siemens' larger SWT-3.6-107 is a 3.6 mW giant with variable rotor speed providing increased aerodynamic efficiency. The gearbox has a three-stage planetary helical design. It is small and relatively quiet. In the future, offshore siting for wind farms will probably become more common, and Siemens' design contemplates this shift. The fiberglass reinforced rotor blades have no glue joints, so they are less vulnerable to lightning and water entry—critical considerations for offshore siting.

As for troubleshooting and repair, these and other utility-scale models are suitable for connection to the Internet, so that they can be remotely monitored. Operating conditions are reported via a conventional Internet connection to the monitoring site, using a SCADA-based graphic interface. Workers can keep track of the electromechanical status including, among other items of interest, voltage, current, frequency, power factor, generator rpm, lubricant temperature, and other variables. If an anomaly is detected, repair crews can be dispatched with the right tools and spare parts before there is a catastrophic event that would be far more costly to repair.

With a record of the reported data in hand, troubleshooting is simplified, although it must be stated that a vast amount of knowledge and expertise is required to maintain and repair this type of equipment.

16

Oscilloscopes Revisited

We have mentioned some of the applications for oscilloscopes. They are used by the more advanced electronic technicians for servicing audio and video equipment, analyzing power quality problems in industrial facilities, and anywhere it is useful to see the shape of a waveform with all its attributes as opposed to merely measuring the voltage.

An oscilloscope (see Fig. 16-1) displays voltage as a function of time. Specifically, there is a horizontal x-axis. Points along this line, left to right, represent the passage of time. Points along the vertical y-axis correspond to various voltage levels. Sometimes intensity is considered a third "perpendicular" axis, and it can be defined to correspond to some other attribute of the signal. The point at which the axes intersect is called the origin.

If the horizontal x-axis were indefinitely long, there would be no problem depicting an electrical signal, either periodic or otherwise, in real time. However, due to the constrained screen width and horizontal deflection voltage requirements, the horizontal x-axis will be limited in length. Accordingly, it is necessary for the beam of electrons to back up and start over at the left side of the screen when they run into the wall at the right side. One problem this makes is that the retrace would make a bright line across the screen, interfering with the display. This difficulty is easily solved by interrupting or blanking out the electron beam during the retrace interval, which is similar to the TV protocol.

A second problem that was encountered in the early stages of oscilloscope development was that when the trace began anew at the left side of the display, the event would rarely coincide with the same point in the waveform. The result would be at best a rapid progression of the waveforms across the screen, making it difficult to interpret, and more likely a meaningless blur of light.

There is also a solution to this problem and it is known as "triggering." The horizontal deflection circuitry is keyed to a selected point in the waveform, such as the positive peak, so that all trace cycles are synchronized and superimposed on one another, making for a stable image.

We will go into this in more detail soon, but first here is some historical background information that will provide perspective.

The earliest oscilloscope-like devices, other than hand-drawn solutions, provided a graphic representation of fluctuating electrical energy by inking a line on paper. The paper was mounted on a steadily rotating drum, rotation corresponding to the passage of time. The pen was attached to a mechanical linkage that permitted it to move perpendicular to

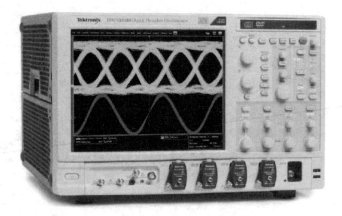

FIGURE 16-1   Tektronix digital oscilloscope with four channels to display and compare wave forms. (*Courtesy of Tektronix.*)

the drum's direction of rotation. This motion corresponded to our *y*-axis. The pen moved in response to electromagnetic fluctuations, the coils wired to the electrical source to be measured. These concepts and the mechanical realization were not difficult for late nineteenth century experimenters, for rotating drums had been used in other contexts. However, actual results were limited by the low speed of the moving pen and by the lack of any triggering mechanism. In order to achieve even a small improvement in resolution, it was necessary to average out the measurements over several hundred consecutive cycles and put them together to create a composite image.

An improvement was the photographic oscillograph. The problem with the paper-drawing device was that the pen and linkage had too great a mass and so it could not respond to even high-frequency waves.

A low-mass device, the moving-coil oscillograph, developed by William Duddell, consisted of a lightweight mirror that was able to move at the higher speeds of the waveforms encountered at the time. Since the photographic images had to be developed before they could be examined, the device had limited practical application.

Further Developments

In the third decade of the twentieth century, a better machine was developed. It consisted of a small mirror mounted on a diaphragm attached to a horn. It permitted images of waveforms up to 10 kHz to be projected on a screen for direct (nonphotographic) viewing. A spinning polygonal mirror generated an unsynchronized time base by reflecting light from an arc lamp.

Various other devices were invented in the first half of the twentieth century, eventually providing higher frequency response, reaching the low audio level. These devices were optical-mechanical constructions and, by their nature, at best there would be limited frequency response due to the inertia inherent in physical as opposed to electronic solutions.

Much higher speeds were achieved when it was realized that the cathode ray tube (CRT) could be used to graphically depict electrical waveforms. The CRT had been developed in the late 1800s. In 1898, about the time of the first edition of the National Electrical Code, Karl Ferdinand Braun built a CRT-based oscilloscope. The signal was applied to vertical deflection plates. Since there was no internal horizontal deflection that would provide timing, this had to be generated by an external rotating mirror. Jonathon Zenneck added an internal time base to the CRT in 1899, making the first swept-trace oscilloscope.

Early CRTs required an electrical supply as high as 30,000 volts to produce a beam that would move across the phosphor face. With development of the heated cathode, the operating voltage dropped to under 1000 volts, making the CRT far more user-friendly.

The first mass-produced, nonlaboratory oscilloscope was built by General Radio, based on the work of V.K. Zworykin, who had developed a permanently sealed CRT in 1931. It had a higher vacuum, making for greater longevity, and a thermionic emitter.

Great advances were made during World War II in England, when innovative technologies were required for servicing the new radar equipment. This was made possible by the development, in the late 1930s, of the first dual-beam oscilloscope by a British firm that was eventually bought out by Raytheon. A great breakthrough occurred in 1946 when Tektronix introduced Model 511, the first triggered-sweep oscilloscope, invented by Howard Vollum and Jack Murdock. Previously, horizontal deflection was governed by a free-running sawtooth wave, generated locally. The sweep had to be synchronized with the signal to be observed, but there were severe limitations in this arrangement.

The repeating waveform is displayed as a stationary graphic image on the oscilloscope when there is triggering circuitry. This means that rise time, phase, and frequency may be measured.

There are a number of different triggering arrangements. Besides initiating a new sweep every time a threshold voltage level is exceeded at the vertical deflection plates, it is possible, for example, to activate triggering based on the polarity (positive or negative going slope) of the signal that is measured.

Improved Scopes

In recent years, new triggering methods have improved the capability of oscilloscopes. Trigger hold-off is one such example. It prevents the oscilloscope from retriggering during a specified time so that there will not be false triggering due to multiple edges in the waveform.

Howard Vollum and Jack Murdock established Tektronix, then and now a leader in oscilloscope development and the first manufacturer of calibrated oscilloscopes as well as multiple-trace instruments that displayed signals simultaneously so they could be compared. This was accomplished by means of time multiplexing or by installing more than a single gun in the CRT.

Other developments, many by Tektronix, enhanced the usefulness of this new instrument, and it began to see widespread use outside the laboratory, particularly in the new field of TV servicing. TVs were sweeping the nation. A new generation of electronic technicians, returning from service in World War II, brought new skills and education into the world of consumer electronics. This is how it often works—social change and technological innovation occurring simultaneously and closely related. Still, the oscilloscope was an analog machine, and as long as this remained true, there would be inherent limitations.

Walter LeCroy, founder of LeCroy Corporation, located in New York, invented the digital oscilloscope, which has dominated the field since the 1980s. The digital oscilloscope differs markedly from its analog ancestor. A high-speed analog to digital converter (ADC) in conjunction with memory microchips records and displays a digital image of the waveform that is being examined. This method is superior to the analog procedure because there is greater opportunity for processing and display. The digital oscilloscope goes beyond the analog instrument in that it can display portions of a waveform that precede the instant that triggering occurs. This is useful in troubleshooting because intermittent and rare anomalies can be captured and preserved electronically for analysis and reference at a future time. Any serious technician will want a digital instrument with as many features as possible.

An electronic technician who is still in an early learning mode and first confronts an oscilloscope is likely to stare blankly at it and think, "How did I get myself into this?" All these knobs with cryptic names like bandwidth enhancement! Nevertheless, be assured that with a little patience and persistence you will get through this and soon be as comfortable using an oscilloscope as a 12-year-old riding a bike is (see Fig. 16-2).

Notice that front-panel controls are grouped into three (sometimes more) sections, often with a heavy border separating them. The controls in each section work together to adjust properties of the oscilloscope. Some of these controls are very important and correct positioning of them is necessary for every reading that you will do.

FIGURE 16-2 Tektronix digital oscilloscope for fieldwork. (*Courtesy of Tektronix.*)

The three principle sections are as follows:

- Vertical—This controls the amplification or attenuation of the signal fed into the input. Adjusting the volts/div will alter the height of the display. For best readability, you will want the positive and negative peaks to approach the top and bottom, respectively, of the screen.
- Horizontal—This controls the time base. Adjusting the sec/div will change the speed of the horizontal sweep, so that you can look at a single cycle or two or more cycles in the display.
- Trigger—The trigger level stabilizes the display of a recurring signal, for example by initiating a new sweep when the voltage on the vertical deflection plates reaches a specified level.

The most common vertical controls include:

- Type of termination, which may be 1 megohm or 50 ohm.
- Type of coupling, which may be ac, dc, or ground, which cuts off the signal.
- Bandwidth, which may be limited or enhanced. The advantage in limiting bandwidth is that noise, a high-frequency phenomenon, is suppressed. However, some high-frequency detail is lost.
- Position, which moves the centerline of the display.
- Offset, which adds a dc component in order to adjust the vertical position of the trace.
- Invert, which may be ON or OFF.
- Scale, which may be in fixed steps or variable.

The oscilloscope can trigger on other sources than the signal that is being displayed. It can trigger on any of the input channels. Additionally, it can trigger on an external source that is connected to the oscilloscope. Alternatively, it can trigger on the ac power source. Another option is to obtain triggering from an internal signal. The general practice is to allow the oscilloscope to trigger on the signal that is fed to the vertical deflection plates.

Probes

Separate from the main oscilloscope chassis but very much a functional part of it is the accompanying set of probes. The probe chosen for a given measurement must match both the oscilloscope and the device or circuit being tested. The probe is part of the circuit, and at high frequencies it possesses sufficient capacitance and inductance to affect the measurement. It also has a certain resistance. It is essential to minimize loading, and to do this the correct probe must be chosen for each task.

For most work, passive probes are used. They intentionally attenuate or reduce the signal. The attenuation factor may be 10X or more. (If it is attenuation, the X comes after the factor. If it is amplification the X comes before the factor, as in X10.)

Untrained personnel believe that to move a telephone to another location in the same room you should purchase couplings and add line cords. It is much better to use concealed telephone wire and put in additional jacks as needed.

Before actually operating an oscilloscope, we should review a few safety considerations that have to do with the lab or shop environment, the instrument itself, and the equipment being worked on. There are two areas of concern. First, do not create a situation where you, a coworker, or a member of the public could be exposed to electric shock. Second, in working on equipment that is to be repaired and put back in service, do not make any modifications or additions that could create a fire hazard. Current-carrying conductors should not be downsized. Any wire with high-temperature insulation must be replaced in kind. If a fuse has blown, find out why and correct the underlying fault, as opposed to just putting in a new fuse and hoping for the best. Replacement fuses must have the correct current rating. Never disable a fuse by shunting around it. In making a repair, do not introduce excess flammable material into the enclosure or do anything that will interfere with air circulation and heat dissipation.

The shop should be clean and well lit, and the bench uncluttered. Flammable solvents, spray paint, toxic chemicals, lubricants, propane gas canisters, and so on should be stored in a locked metal cabinet that is remote from the work area. The bench should have a nonconductive, fireproof surface. The entire area including the floor should be dry and free of debris. A dry rubber mat on the floor will lessen the possibility of severe shock.

Verify that all receptacles are grounded and, where required, are of the GFCI type. There should be plenty of receptacles so that there will not be a proliferation of extension cords and multipliers. The receptacles should be on two separate branch circuits that are not the same as the lighting.

Do not work on TVs or other equipment that may contain high voltages without taking special precautions. Remember that dangerous electrical energy is retained within the enclosure long after the equipment has been powered down and unplugged. This is because inside there are capacitors that are capable of holding a charge for a long time. Do not work inside the enclosure unless you know how to identify the wiring and components where high voltages linger, and that you know how to bleed them out. Even then, as an added precaution, do not touch these surfaces. Open chasses that may harbor high voltages should not be left unattended on the bench.

Check that the oscilloscope is grounded. Sometimes in the past, ground plugs have been cut off so that the cords can be plugged into obsolete nongrounding receptacles. Verify that this is not the case and that there is continuity between the grounding bar in the distribution box and the oscilloscope case if it is metal. (Some equipment with cases made of nonconductive material and no exposed metal parts are not required to be grounded and do not have ground prongs.)

It is also necessary that the oscilloscope be properly grounded in order to take accurate measurements. The instrument must share the same ground as the equipment being tested. Otherwise, the difference in ground potentials will render the oscilloscope readings meaningless. Do not do anything foolish like running a wire to a nearby water pipe to pick up the ground connection.

Additionally, if working with integrated circuits, ground yourself. This is to prevent small static charges from damaging the semiconductors. A good way to prevent static buildup is to wear a grounding bracelet, which is plugged into a good solid ground. Be sure to take off the grounding bracelet if you are going to be working on anything that might be energized.

Plug in the oscilloscope and read the markings on all the controls. Locate the input sockets. If there are AUTOSET or DEFAULT buttons, use them to automatically set up the oscilloscope for taking a reading. If the oscilloscope does not have these controls, the setup will have to be done manually. The usual procedures are as follows:

- Activate Channel 1.
- Place the vertical volts/div and position controls at midrange.
- Disable variable volts/div.
- Turn the magnification controls to the OFF position.
- Switch Channel 1 input coupling to DC.
- Turn trigger mode to AUTO.
- Turn trigger source to Channel 1.
- Switch off the HOLDOFF control.
- Place horizontal time/div and position to midrange.
- Adjust Channel 1 volts/div so that the signal will fill a good part of the screen. If there is clipping or distortion, the control can be retarded later.

Acquiring a Good Display

The next step is to compensate the probe to the oscilloscope, if it is a passive attenuation probe. Modern oscilloscopes have a square wave reference signal generated within. There should be a terminal on the front panel where it is available. Connect the probe to Channel 1, making sure that channel is activated. Attach the ground clip of the probe to the ground. Touch the probe tip to the square wave terminal. You will see the square wave displayed on the screen. Adjust the probe controls until the corners of the square wave are square. Rounded corners are an indication that the probe is improperly adjusted and loading the circuit. Having progressed to this point, you should be able to say why a probe that is out of adjustment will cause the square wave to exhibit rounded corners in the display. (Hint: It has to do with reactance.)

In order to measure voltage with an oscilloscope, get a good display of the waveform on the screen. Then, count the divisions between the points on the waveform that interest you. In this way, you can determine peak-to-peak, peak-to-ground, root-mean-square (RMS), and other values. A better way, if available in your oscilloscope, is to use the cursor so that you do not have to count graticule marks. You can position two cursor lines to enclose the part of the waveform that you want to measure and the readout will show the voltage. This will provide more accurate data than can be estimated visually.

The time measurement, taken by looking at the display or bracketing with the cursor, gives a direct quantification of the period of the waveform in seconds, usually a small fraction of a second. You can take the reciprocal of this quantity to find the frequency (see Fig. 16-3).

Often, you will want to know the characteristics of a pulse. If a digital circuit is not working as it should, it may be because the pulse shape has become distorted. In addition, in data transfer the timing of pulses is critical. Defects like these can be plainly seen in the oscilloscope display. The important parameters in a pulse are rise time and pulse width. The rise time is defined as the elapsed time between 10 percent and 90 percent of the peak voltage. Pulse width is defined as elapsed time between two successive pulse lows. This is generally measured at the 50 percent level so that the figure is not influenced by distortion at the extremes.

Pulse measurements are important aspects of the troubleshooting process in digital data systems, and they require a bit of expertise with the oscilloscope. What is needed is some savvy with trigger holdoff and familiarity with capturing pretrigger data. Another essential area is horizontal magnification, often necessary when dealing with fast pulses.

Phase shift measurements involve a palpable new level in oscilloscope expertise. If you aspire to be an upwardly mobile electronic technician, here is your chance. What is phase shift? It is defined as the difference in position along a time line between two signals that

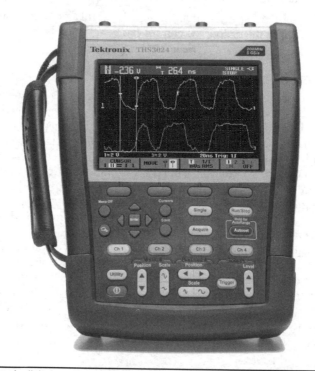

FIGURE 16-3   Tektronix digital oscilloscope with cursor to define portion of waveform to be measured. (*Courtesy of Tektronix.*)

are otherwise identical. The best way to compare two such signals is to graphically represent them in a single combined image by feeding one of them into the vertical deflection circuit in the usual manner, and the other into the horizontal deflection circuit, which is not usual. In other words, one signal is being timed by the other signal. This is called the *XY* measurement because both axes are creating a single voltage trace. The image that we see on the screen of the oscilloscope is not at all like the waveforms we have encountered up to this point. They are called Lissajous patterns. They are named after the French researcher Jules Antoine Lissajous (1822–1880) who conducted experiments on the combinations of oscillations, not in an electrical context, but produced by vibrating tuning forks attached to mirrors perpendicular to one another. These same patterns are displayed on the screen of an oscilloscope with the two signals connected as described. The patterns differ depending upon the amount of phase shift and the ratio of the *X* and *Y* frequencies.

In troubleshooting, an oscilloscope has a great many applications, and it is capable of providing insights far beyond those obtained with a multimeter. In servicing audio, video, and many types of industrial equipment, many schematics include graphics that show oscilloscope displays when the equipment is operating normally. Assuming that the power supply has been found to be performing properly and that speakers and picture tubes or other output devices are judged to be not at fault, an appropriate procedure is to follow the signal from stage to stage using an oscilloscope, and to determine where it deviates from the graphic in the schematic. Then, again referring to the schematic, you can look at the outputs of individual components and decide which ones should be replaced.

Another oscilloscope application is to check out a VFD, when a motor is performing poorly or as part of a preventive maintenance program. The underlying fact about VFDs is that they generate pulses with very fast rise times. This is the power that is delivered to the motor. With the fast leading edges, the voltages in many ways act like high-frequency electrical energy and this can cause problems. If there is an anomaly in the drive, cable, motor, or intermediate connections, there can be harmful reflections that will make for abnormal peaks in the waveform. Connect the oscilloscope successively to the motor terminations and look at phase-to-phase and phase-to-ground waveforms. Abnormally high voltage transient peaks, above the maximum motor rating, will stress the motor insulation and cause motor failure soon if this has not already happened. It is suggested that these readings be taken as part of a preventive maintenance program and the results saved so trends can be spotted early before damage is done.

What Lies Ahead?

A look at early 1880s outdoor photographs of Manhattan streets reveals a vast proliferation of aerial power lines. Wires are everywhere. Wooden poles with multiple cross arms bearing glass insulators delivered power and telephone service to commercial and residential buildings, with stock ticker, telegraph, and other conductors strung along for good measure. Somewhere along the line, it was perceived that to continue like that would be disastrous. Eventually, even the most skilled birds would not be able to fly through the air. The utilities resisted burying lines because of the great expense. The great blizzard of 1888 dumped over 20 in. of heavy snow that stuck to the lines, brought wires to the ground, leaving New York City paralyzed for weeks. Municipal authorities ordered all lines buried immediately and when the utilities refused to comply, city workers commenced cutting down the poles. By the end of the following year, after court battles, aerial lines in the region were a thing of the past. Although it is more expensive, in increasing numbers of cities and suburban communities, utilities are burying their power and communication lines, and high-voltage transmission lines increasingly are going underground and beneath bodies of water.

The advantages go beyond the aesthetic. Underground lines are invulnerable to catastrophic weather events and texting or intoxicated motorists. Lightning damage can occur, it is true, but advanced protection devices are minimizing the problem. When repair becomes necessary, accurate mapping and improved excavating techniques make the repairs less costly. The correct approach would seem to be burial of all new work and incremental burial of existing aerial transmission and distribution lines.

This, plus wind farm and localized solar PV installations, will make for great changes in the way electrical work is organized. Additionally, changes in cyber equipment and data speed expectations including expanded use of optical fiber will make for increased demand for electrical workers at all levels of the profession. It is possible that in the future electricians, defined broadly, will become what truck drivers are today—the largest single work category in the world.

All these changes will mean new imperatives for those of us in the profession. We will have to know more and have greater expertise to continue in our jobs. The future will bring enormous change, but it will come on gradually and there will be time to adopt.

Throughout, this book has taken the position that in the field of electrical work, knowledge is the most important tool. We should always be inquiring into new areas, new technologies, and new ways of thinking about everyday things. It is highly unlikely most of us will be called upon to troubleshoot and repair a particle collider or the International Space Station, much less disassemble a discarded one of these to see how it works. Nevertheless, as a thought experiment, we should take one apart in our imagination in the interest of gaining insight into some of the more mundane corners of the universe in which we do our business.

Certainly, there is at best a limited connection between quantum mechanics and repairing a lamp. However, when there is data flicker in a large computer network, all kinds of unusual concepts will become relevant.

If you can fix an electric drill, you can refurbish a particle collider. It will take longer, maybe more than a lifetime, require years of background research just to get started, and the stakes are much higher. However, the same basic troubleshooting techniques apply.
The more you understand the equipment in question, the easier it will be to make the repair. If you do not understand the equipment, look it over for visual clues. If that does not work, go back to the Internet or the printed word and see what you can discover.

A particle collider differs from and goes way beyond a particle accelerator. Before the flat-screen era, nearly everyone had at least one particle accelerator in the form of a CRT TV or computer monitor. The defining element in a CRT is the electron gun, situated at the small backend adjacent to the input pins. Its function is to emit electrons, which are then accelerated to a very high speed, deflected in a scanning pattern, and varied in intensity. When these electrons hit the phosphor screen, dots of light are produced, forming the picture.

To initiate the electron beam, the gun has a heater, similar to a light bulb filament. The cathode, thus heated, emits electrons. The accelerating anode gathers up these electrons and propels them in the direction of the screen. The focusing anode converts the flowing electrons into a narrow beam so that the bright spots that appear on the screen will be small enough to create a picture with satisfactory resolution.

In a black and white TV, there is a single electron gun. A color TV has three electron guns, each producing a separate stream of electrons that, at the screen, energize different phosphor coatings corresponding to the three primary colors used in the color TV protocol: red, green, and blue. To make it work, the beam has to scan the screen, successive horizontal lines progressing vertically at a prescribed speed.

In an oscilloscope, the beam of electrons is made to scan by means of deflection plates, the signal being fed to the vertical deflection plates and the time base fed to the horizontal plates.

A nonflat screen TV works in the same way, but instead of deflection plates there are deflection coils. The question arises—can a television be made into a home-brewed oscilloscope? The answer is yes. It involves going inside (watch out for lethal voltages even after the TV has been powered down) and disconnecting the horizontal and vertical deflection coil leads, extending them outside the cabinet, and feeding the appropriate signals to them. A computer can also be adopted to display electrical waveforms. The sound card serves as an ADC (analog to digital converter). Appropriate software is available from a number of sources. Neither of these expedients provides anything approaching a full-featured digital storage-triggering oscilloscope with an internal reference voltage source.

Probing the Universe

A particle collider differs from the particle accelerator described previously, but it uses some of the same basic concepts. A particle collider accelerates particles, primarily protons, to extremely high energy levels and causes them to collide. Then, by means of detectors, it captures the results of these collisions so that they may be studied by researchers.

The largest particle collider in the world is operated by the European Organization for Nuclear Research (CERN). It is headquartered outside Geneva, Switzerland, and has almost 4000 workers and approximately 10,000 visiting research scientists and engineers. At present, the primary focus of interest within CERN is the Large Hadron Collider (LHC). It is located in a tunnel over 300 ft underground, having a circumference of more than 30 mi. The tunnel itself was recycled from a previous project that was discontinued in 2000. The tunnel straddles an international border so that part of it is in France.

In the tunnel, two types of particles are accelerated and made to collide. They are protons and lead ions. Earlier particle colliders fired particles at stationary targets, but the current procedure is to fire two beams at a common point so that they collide head on. This provides double the energy.

The object behind all of this is to replicate conditions just after the Big Bang, when all potential matter and energy were concentrated at a very small point. For reasons that are not fully understood, this point exploded to form the entire universe.

How do we know this is true? In the twentieth century, observation of spectral shifts in the light from distant stars led astronomers to conclude that these bodies are moving apart rapidly. In other words, the universe is expanding. This seems to imply that it originated at a central point. Subsequent observations and theoretical reasoning have reinforced this view. The original event has been given the picturesque name "the Big Bang." By replicating the conditions immediately following this event, it is believed that we can uncover information concerning nature and relationships among the many elementary particles that are known to exist. At the LHC facility at CERN, the two proton beams following the circular track that is in the tunnel each have an energy of 7 TeV, so when they collide the energy is 14 TeV. This value is far higher than achieved previously.

In all particle accelerators including that familiar CRT, particles move through a vacuum and are accelerated by means of magnets. Among the features that make the CERN facility unique are the length of the circular track, the high purity of the vacuum (to prevent unwanted collisions with gas molecules that would impede the beam), and the number and power of the magnets that accelerate and control the beam. In addition to the magnets, there are electronic cavity resonators operating at microwave frequencies, which further refine beam characteristics.

There are about 9600 magnets in the LHC. The purpose of the cavities is to keep the protons tightly bunched in order to facilitate a large number of collisions. For the magnets to produce the very great amount of magnetic flux required, it is necessary for the conductors to have as small a voltage drop as possible. This is achieved by cooling them down to close to absolute zero. The magnets have niobium-titanium cables. They become superconducting at a temperature of 10 K, conducting electricity with no resistance.

The current that flows through the dipole windings is 11,850 amperes. The windings consist of a cable made up of 36 strands, each strand formed from up to 9000 substrands. In all, there are almost 5000 miles of cable in the tunnel, with numerous terminations. If any one of these terminations is faulty, it will heat up and fail catastrophically, as happened in October 2008, causing a 1-year shutdown for electrical repairs.

To make possible this enormous super-cooled project, helium was chosen as the refrigerant. Helium gas, at atmospheric pressure, becomes a liquid at 4.2 K. At 2.17 K, there is another phase change and the helium becomes what is known as a "super fluid." In this form, the helium has very high thermal conductivity, making it an extremely efficient coolant and the best choice for refrigerating superconductors on the scale necessary at CERN's LHC.

Each beam in the LHC is made up of 3000 bunches of particles. There are 100 billion particles per bunch. When the beams cross, there are only about 20 collisions because the particles are so small, with relatively vast amounts of space between them. However, bunches cross about 30 million times per second, so in the LHC there are as many as 600 million particle collisions per second. The beams circulate for several hours, traveling approximately 13 billion miles. The protons move at close to the speed of light, making over 11,000 circuits per second.

There are currently six experiments being conducted at the LHC:

- ALICE is engaged in the study of lead-ion collisions. The object is to discover the properties of quarks and gluons where at extremely high pressure and density, they are not confined inside hadrons. It is theorized that this was the state of affairs shortly after the Big Bang, prior to formation of protons and neutrons.

- ATLAS is the largest collider detector ever built. It consists of eight superconducting magnets over 80 ft long. The purpose is to verify the existence of the Higgs boson, supersymmetry, and extra dimensions.

- CMS is pursuing the same theoretical program as ATLAS, but using different methods. It consists of a very large superconducting solenoid that is a cylindrical coil of superconducting cable that will generate a magnetic field 100,000 times greater than the Earth's magnetic field.

- LHCb attempts to discover the meaning of the asymmetry between matter and antimatter that occurs in interactions of B-particles, which are particles containing the b quark. Instead of a single detector, there are a number of subdetectors.

- LHCf is a small experiment consisting of two detectors. The purpose is to gain information about the nature of high-energy cosmic rays.

- TOTEM is an experiment that is designed to measure the exact size of a proton. The detectors are housed in vacuum chambers placed near the collision point of the CMS experiment.

Tracking devices that "see" the collisions take either of two forms. They can be gaseous chambers in which a gas is ionized. Electrodes that are under very strong electric fields pick up ions or electrons. The other type of tracking device is a semiconductor detector. The particle creates electrons and holes as it passes through a silicon reverse-biased semiconductor.

The LHC power consumption is approximately 120 mW. A good part of this electricity is needed to run the refrigeration equipment. Because of the superconductivity, energy to power the magnets is much less than it would be otherwise, even when deducting the amount of energy required for cooling.

Human eyes do not see the collisions. What happens is that a great many sensors accumulate data, which is stored and processed in a computer network. There is far too much data for any one computer to store and analyze.

The LHC at full operation produces 15 million gigabytes of data per year. CERN has assembled a distributed computing and data storage infrastructure known as "The Worldwide LHC Computing Grid." It is the world's largest computing grid and allows researchers to access data in real time as it emerges. One advantage is that there are no single points of failure. Moreover, in the future, in can be reconfigured to support new research projects as they come online.

CERN was also instrumental in developing the Web, a subdivision of the Internet whereby individual computers can access material in servers by means of hypertext. Tim Berners-Lee, a physicist, proposed the system in 1989. The first website and server combination was set up and it expanded rapidly in the years following.

We have looked at some of the details of CERN's particle collider, an enormously complex electrical installation, in order to provide some perspective and to suggest the fact that troubleshooting and repairs are relevant in any size project.

Leaving Earth

An equally challenging undertaking has been the International Space Station. Like the CERN project, it was constructed in the 1990s and the following decades, and at the heart of each is a sophisticated electrical system whose principle requirement is reliability.

Most spacecraft have electrical systems that operate at 28 volts dc. The Russian section of the International Space Station follows that pattern. The American section operates at 124.5 volts dc. DC step-up and step-down converters allow for power transfer between the two sections.

The International Space Station's electrical loads are supplied by a solar PV system. The station orbits the earth once every 90 minutes, and for over a third of this time, the station is in the Earth's shadow and there is no PV output. Since the PV system is stand-alone, there is battery backup. The energy storage system consists of nickel-hydrogen batteries.

The PV output ranges from 137 to 173 volts dc. This is reduced and regulated to the 123 to 126 volts dc system in the American section. Having a voltage level this high mitigates voltage drop. (The International Space Station is the size of a football field with 8 miles of wire distributing power.)

There are 400 solar cells connected in series in each string, and 82 strings connected in parallel to create the current capability required to power concurrent loads. The cells are silicon with a 14.5 percent efficiency. They are attached to a printed circuit underlayment that conveys power downstream. There is a bypass diode for each group of eight cells.

The arrays are of the tracking type for greater output. Two perpendicular axes facilitate accurate pointing toward the sun. Power from the solar PV system is fed via eight independent power channels to four main bus-switching units. Ordinarily, each power channel supplies a different set of loads within the International Space Station. If a channel fails, a main bus-switching unit can feed power to the affected loads through an alternate channel. All electrical functions are computer controlled and can be manipulated either by the International Space Station crew or by ground personnel.

DC-to-DC converters supply regulated power at the 124.5-volt level to remote power controller modules. They enable remote switching for loads and overcurrent protection. One of these boxes feeds the Russian section through a dc-to-dc converter that steps down to 28 volts.

The primary system, from solar cells to dc-to-dc conversion equipment, and the secondary system, from that point downstream to the utilization loads, have quite different configurations. In the primary system, the power channels ultimately connect to specific loads. In the event of a fault, these channels can be switched around to ensure that critical loads remain powered up. The secondary system, operating at 24.5 volts dc (nominal) does not have this cross-strapping capability. Instead, other measures that provide a high degree of reliability are in place. Components can be wired with multiple power input sources. In many cases, more than a single component serves the same purpose. Thus, there is redundancy not in the secondary electrical system, but in the end-use equipment. Also, there is plenty of extra wire on board, so crew members can make jumpers and improvise connections if need be.

Suppose you were a crew member, the best electrician aboard, and you were suddenly faced with a life-threatening malfunction. It comes to you that you are totally in over your head. What recourse would you have? The answer is that you could radio ground control and an engineer could certainly walk you through to a solution. In all troubleshooting and repair operations, especially in today's Internet and tech help lines environment, there is ample opportunity to obtain outside assistance, and this is often the answer to a perplexing problem, where complexities extend beyond one's immediate knowledge base.

The International Space Station electrical system, like most of its earthly counterparts, includes power surge and overcurrent protection at various points between the solar cells and end-user loads. There are voltage, current, and temperature sensors in almost every piece of equipment, with firmware embedded in the equipment and installed in computers. If any variable goes abnormal, appropriate action is taken automatically. If there is overcurrent, each successive device going in a downstream direction is more sensitive (set to cut out at a lower current level) and quicker acting. This is so that a fault will not disable upstream portions of the circuit. Also, if there is a surge, affected segments are quickly isolated so as not to imperil other equipment. Additionally, power production is stopped when array output voltage drops below a specified value, so that PV cells do not continue to power a fault.

This short survey of the CERN and International Space Station facilities should serve to place some of our daily maintenance and repair activities in perspective. In the interest of opening a further thought-provoking topic, for those who have not already considered it, we shall now provide an overview of laser technology and fiber optics.

Commonly, electromagnetic radiation or sound propagates equally in all directions from its point of origin. Imagine a single pulse occurring at a certain location in space. The advancing energy front takes the form of a spherical shell, the radius increasing with the passage of time. Since the area of a sphere varies with the square of the radius, the strength of the signal diminishes with the passage of time and the distance from the origin according to that relationship. Attenuation is a consequence of that spatial dilution, not of any kind of resistance in the space that is traversed.

Electromagnetic radiation can be made to travel in a narrow beam by focusing it with a lens, reflecting it at a parabolic mirror, or using a laser to create a narrow beam of coherent light, which spreads only a small amount. Another way to constrain electromagnetic radiation to a single narrow path so that it is not attenuated as the distance from the source increases is to confine it to a waveguide. This may be a natural phenomenon or by means of a manufactured device. The waveguide varies in size according to the wavelength and it varies in shape according to characteristics of the wave and how it is to be handled. For a radio telescope, the waveguide is massive and has to be installed using a crane. Alternatively, it may be so small that it has to be handled with tweezers in order to affix it to a printed circuit board.

Waves remain inside the waveguide and lose no power while propagating because they are totally reflected from the inside surface of the waveguide wall. The wave zigzags from side to side, all the while progressing the length of the waveguide. For this to work, the waveguide must have a dimension that is the same order of magnitude as the wavelength of the signal to be conveyed.

As frequency increases, waveguides pick up where coaxial cable leaves off. Coax becomes untenable at high microwave frequencies due to reactive losses. Moreover, waveguides are impractical at lower frequencies because they would have to be very large.

As we saw in Chap. 10, a satellite dish antenna makes use of a waveguide to convey the electromagnetic radiation to the LNB where a local oscillator and mixer drop the frequency sufficiently so that the signal can be brought into the building to the receiver.

Coherent Light

LASER is an acronym for Light Amplification by Stimulated Emission of Radiation. A laser is a piece of equipment that takes advantage of this process to produce a beam of light that has some unique and very useful characteristics. First, laser light is coherent. This means that at each point in time the light maintains the same phase relationship. Second, the light in the beam is monochromatic. Since it is confined to a single wavelength, it appears as a single color of great purity. Finally, the light is collimated so that the beam is narrow and spreads very little as it travels through space. A beam of nonlaser light, even though it may be focused initially in a single narrow beam, inevitably spreads as it travels away from the source. This is because it interferes with itself, causing the photons to disperse.

The properties of a laser-generated light beam make it useful in a great many diverse applications. Most advanced carpenters, for example, possess a laser level or transit. Simple versions of this instrument permit the user to project a small, precise, and highly stable spot of light on a work surface 100 ft or more distant. By means of a horizontal adjustment and bubble vial, the beam can be made to travel in a level line, so it is useful for setting grades on a construction site. Indeed, electricians use the same instrument for such diverse tasks as setting light fixtures along an accurate reference line or aligning long runs of conduit.

A similar setup aids in positioning logs accurately with respect to the blade in a sawmill. On a smaller scale, the laser indicating line or spot is useful for aligning the work piece on a table saw or chop saw. There are countless additional applications in our everyday world—supermarket barcode reader, DVD read and burn device, computer hard drive, laser pointer, medical and dental machinery, and others. The electrician or electronic technician will be called upon to diagnose and repair this sort of equipment. To get started, it is necessary to know how it works. Before entering into this discussion, we have to emphasize some safety warnings. For even low-powered devices such as the laser pointer, do not look directly toward the beam or its reflection. Permanent damage to the retina can result. Do not be fooled by the dark glasses that are supplied with some laser levels. Their purpose is to make the laser spot or line more visible in full outdoor light by filtering out all but the laser wavelength. Consequently, the pupil of the eye opens wider, making the retina actually more vulnerable to laser damage. If you have occasion to work around more powerful equipment such as that used in metal-cutting operations, it is necessary to keep clear of the beam to prevent serious injury. When working on this equipment, power it down and observe lockout procedures. Also, beware of power supply and laser pump capacitors that can store dangerous amounts of electrical energy long after the machine has been disconnected from the power source.

To understand how the laser is able to create its unique beam of light, we have to consider the inner workings. A laser somewhat resembles a transistor in that an external source of power must be supplied. The amplification process in a laser as well as in a transistor is not some magical process that creates a greater amount of energy from within, but it depends upon a continuous or pulsed supply of light or electricity to facilitate the process of light amplification. To understand how this works, we shall look at a simpler, nonlaser process known as thermal emission.

Electrons travel around the nucleus of the atoms with which they are associated. (Some electrons are separated from the atom, but that is another story.) These atoms do not orbit in a single plane, as do the planets that revolve around our sun. Instead, their orbits vary longitudinally, so they may best be visualized as following paths that comprise three-dimensional shells surrounding the nucleus. Moreover, the orbiting electrons are not spaced continuously at all distances from the center. Instead, the orbits occupy discrete isolated zones that in comparison to their sizes are vast distances apart, if we can conceive of such a thing. The greater the distance from the nucleus, the higher the energy level of the electron. This all-important fact and key concept makes possible the laser action. Before describing how this works, we will look at "ordinary" thermal radiation so that we can gain perspective on the laser mechanism.

If a body of material in a vacuum is heated sufficiently, it will emit electrons. Since they carry a negative charge, the body acquires a positive charge that increases as more electrons depart. (If a battery is connected to the body, that charge can be neutralized and the charge of the body kept at its original level.)

Disorganized Light

When heat is applied to a body, it will also emit photons, which are seen as visible light. If you heat a piece of iron with a torch, as the temperature rises you will first see a faint red glow that becomes brighter as the temperature increases so that eventually the metal radiates photons of all wavelengths, and the perceived color approaches white. This emitted light is a continuous, nonpulsed, disorganized form of radiation made of a mix of wavelengths not in phase and radiating in all directions. It diminishes in intensity with the square of the distance from the source. The light is useful for illumination and it makes phenomena such as photosynthesis and solar PV power possible. However, this light is not like laser-generated light because it is not coherent, monochromatic, or collimated. These attributes belong to laser-generated light. To create the laser effect, a specialized piece of equipment with an external power supply is needed. The laser makes use of the quantum processes of absorption and stimulated emission. The fact to remember is that in matter, the electrons are at a higher energy state when they inhabit an orbit (shell) that is farther from the nucleus and they are at a lower energy state when they inhabit an orbit (shell) that is nearer to the nucleus. Electrons can spontaneously move from the higher to the lower state. If they do so, they emit a photon, or particle of light.

If an electron is in the lower energy state, closer to the nucleus, and if it is struck by a photon, the electron will absorb the photon and simultaneously jump to a higher energy state, that is, move to the orbit that is farther from the nucleus. The electron moves back and forth between these energy states, each time absorbing or emitting a photon. This process can be exploited to create a coherent, monochromatic, and collimated beam of light. The essential parts of a laser are a power source (light or electrical), a pump, a resonant cavity in which the amplification process takes place, and two mirrors that are mounted at each end of the cavity. The mirrors must be as near parallel as possible. The back mirror is completely reflective, while the front mirror is 99 percent reflective so that a small amount of light is intentionally allowed to exit from the cavity, forming the laser beam.

The ordinary condition of matter is for the electrons to be in the lower energy state. From time to time, an electron will jump to the higher energy level because of photon absorption, but it will quickly fall back to the lower energy state because of spontaneous or stimulated emission.

For laser action to take place there has to be a condition known as population inversion where electrons move to the higher level and are available to drop back, accompanied by photon emission. A precondition is injection of energy on an ongoing basis. That is where the pump comes in.

A common gas laser is the helium-neon device. Helium can be excited to a level that is very close to neon. When helium and neon atoms collide, the energy is transferred from the helium to the neon. The helium is the pumping medium that makes possible the population inversion so that laser action can proceed. For any laser action to take place there must be a pumping mechanism. There has to be an injection of energy, and this can happen in a variety of ways.

The carbon dioxide laser operates at a much higher power level, in excess of 10 kW continuously and much higher in the pulse mode. Its method of operation is similar to the helium-neon laser. Nitrogen is the pumping gas, receiving its energy from electric discharge. Because of its high power, it is employed in welding and cutting operations.

The argon ion laser is another powerful device, able to output up to 100 watts of continuous power in the form of hot plasma.

The ruby laser, the first such device, was built in 1960. Ruby is actually aluminum oxide that contains a small amount of chromium. It operates in short pulses, receiving a pulse from the flash tube. A laser light beam is emitted during the period of time when there are excited atoms in the ruby rod. Upon excitation, the electrons go first to the highest of three levels. Then they drop to an intermediate level without emitting light or any other radiation. Finally, they drop to the lowest level, this time emitting the usable laser output. The pumping is optical, accomplished by means of a flash tube, with an electrode at each end, coiled about the ruby rod. This illustrates the fact that various laser types have differing methods for injecting the pumped energy, but in all cases there must be an external source of energy that is pumped into the lasing medium.

The neodymium-YAG (yttrium-aluminum-garnet) laser is a solid-state (not to be confused with semiconductor) device. The Nd^{3+} ion serves to dope the YAG material in which population inversion takes place. Here again there is a flash tube with two electrodes. These lasers are very powerful, capable of producing over 1 kW of continuous power and much higher levels in the pulse mode.

Neodymium-glass lasers achieve even higher power because the pulse is very short, in the order of 10^{-12} sec with power reaching 10^9 kW. They are used in research to trigger thermonuclear fusion.

Laser diodes are constructed by joining two layers of doped gallium arsenide. The p-n junction is forward biased. At lower bias levels, the device functions as a conventional LED, but with increased current laser action. The photons bounce back and forth between two mirrored surfaces, one of which is partially reflective so that a coherent beam of light can emerge.

Dye lasers are tunable, meaning that they can be made to lase over a wide range of frequencies. The dye laser medium is pumped by means of a flash lamp or another laser.

All these laser types have in common a pumping mechanism that injects energy into the medium so that population inversion and lasing are achieved. What makes them work is the fact that light that is amplified by stimulated emission is always coherent, monochromatic, and, if there are two parallel mirrors, collimated.

Because of the large number of types of lasers and the variety of equipment in which they are embedded, troubleshooting and repair procedures of necessity are specific to the unit that is the focus. It is usually obvious, as in the case of a laser computer mouse, if the device is not lasing and this will tell you whether to look at the laser module or the equipment that supports it or makes use of its properties. Of course, the priority should be to check the power supply. To determine if the voltage is correct, you may need the schematic or you can do an Internet search.

When working on a diode laser as seen in DVDs and similar equipment, avoid touching the terminals because the laser is easily damaged by static discharge. Also, do not bring a heated soldering iron close or the laser will never again work.

Semiconductor lasers have peak power immediately after being turned on, and then the intensity of the beam diminishes. That is because the rise in temperature causes pump diodes and crystals to become less efficient for the same reason that a solar PV cell has less output at a higher ambient temperature.

In conjunction with higher-powered lasers, there are multiple interlocks. If the cooling system fails or an enclosure door is not closed completely, power may be interrupted. The other possibility is that neither of these situations has occurred, but the safety switch or sensor has gone bad.

All of this is by way of introduction to the subject of fiber optics, in many ways the wave of the future.

Track lighting is difficult to install, in part because different makes go together altogether differently and the parts are not at all interchangeable. The best approach is to study installation instructions carefully before proceeding or even planning a job. Consult NEC Chap. 4 for circuit sizing.

By the process of total internal reflection, optical fiber transmits light along its axis. It is a dielectric, or nonconducting, waveguide. You have probably skipped a stone across a pond. The rock will actually bounce off the water a number of times before it loses speed and disappears below the surface. This strange behavior is because the stone travels at a very slight angle to the air-water boundary, horizontal inertia carrying it with sufficient force to prevent it from moving down into the water in response to gravity. This is what happens in the fiber optic waveguide. The photons skip along between the inner surfaces of the core-cladding interface, totally reflected rather than escaping into the outer world. If the angle of entry is too acute, the light will not reflect, instead escaping the confines of the fiber so that it will not be conveyed to the far end. For this reason, there is a definite acceptance angle that the entering light must observe if it is to be successfully transmitted the length of the fiber (see Fig. 17-1).

Because of this phenomenon of total internal reflection and because of the nature of light particles in contrast to electrons, optical fiber has a different set of characteristics that stand in contrast to copper conductors as used to conduct electrical current.

Optical fibers are faster, have larger carrying capacity, can transmit greater distances without need for repeaters to amplify the signals at intervals, have greater resistance to RF interference from outside radiation, electric motors, etc., and they require less maintenance.

FIGURE 17-1   Outside fiber optic installation requires a large capital investment. (*Courtesy of Fiber Optics Association.*)

While the cost per foot is greater than for copper conductors, the cost is less if you factor in the greater bandwidth available.

The light is created by LED or injection-laser diode (ILD). It is close to infrared, with a wavelength of 850 nm for short distances and 1300 nm for long distances on multimode fiber, and 1300 nm for single-mode short distances or 1500 nm for single-mode long distances. Single-mode fiber has a diameter of up to 10 mm, making it comparable in thickness to a human hair. Only a single strand is needed for signal propagation. Wave-division multiplexing increases performance.

Multimode fiber provides high bandwidth at high speeds. Since there are numerous paths for the light to travel, there is greater capacity; however, in long runs the multiple paths become offset due to minor propagation variations so that the signal can acquire distortion. For gigabit and higher speeds, single-mode is generally used.

Most outside-plant fiber optics is single mode. It can extend hundreds of miles, be hung on poles, buried underground, in conduit, or underwater. Cables will contain hundreds of fibers, each of which, through the miracle of multiplexing, can carry many signals simultaneously.

Since cables are available only in lengths of fewer than 3 miles, a large number of splices is required for longer runs (see Fig. 17-2). Fusion splicing is the usual method. The process is exacting, but because technicians perform the procedure on a daily basis, it becomes very familiar and within your capabilities.

Indoor or premises cabling is by nature a different trade. Since lengths are short, splicing is not necessary. Most indoor work is done with multimode. Because of the shorter lengths, there is not the distortion problem described previously (see Fig. 17-3).

FIGURE 17-2   Long outside fiber optic runs require precision splices. (*Courtesy of Fiber Optics Association.*)

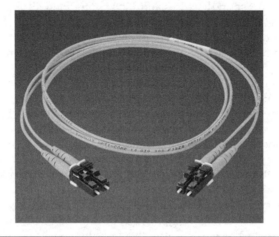

FIGURE 17-3   Fiber optic patch cord. (*Courtesy of Panduit.*)

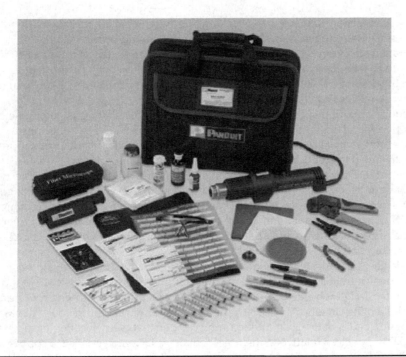

FIGURE 17-4   A variety of specialized tools are needed for working with optical fiber. (*Courtesy of Panduit.*)

Connectors, mostly SC, ST, or LT, are easily installed in a procedure that is not high-tech. Each run must be tested and written records kept. A source and power meter and a flashlight-type tracer are used (see Fig. 17-4). A couple thousand dollars worth of installation and test equipment is all you need. Competent electricians will have no problem completing the coursework and getting into this rewarding line of work.

Finding the Fault

When an installed optical fiber network is found to be deficient, the best way to start is to consult the documentation that was generated during the testing phase following initial installation. This assumes that these records were in fact created and are still available. The original testing should have uncovered any installation flaws, which can include improper pulling technique such as pulling the fiber by the outer jacket rather than by the strength member, using too much force, kinking, excess number of bends, and so on. Other problems include cable damage that occurs during stripping prior to termination or splicing. A nick in the fiber that is barely visible will cause data loss if not total transmission blackout. These problems can be subtle and the manifestations intermittent. For this reason, records should be kept on insertion loss for each fiber. In premises cabling it is common practice to replace the entire segment rather than attempt to repair a damaged run.

Bad terminations are the principle cause of poor performance in premises fiber optic cabling. Using a microscope, examine for damage including dirt or scratches. If enough slack was left at the end, the bad terminations can be remade.

When a connection is found to be working poorly or not at all, it is common practice to switch over to a spare unused fiber. In such cases, this should be noted in the documentation so that in the future a similar change will not attempt to put a bad fiber back in service. In any event, whenever changes are made or inspections performed, it is essential to update the documentation. Print copies should be made because it is likely that the fiber optic network will outlast the computer hard drive.

Test equipment for inside optical fiber work includes the fiber optic power meter, test source, connector adapters for the power meter—ST, SC, and LC—a connector inspection microscope, and a visual tracer and continuity checker or a visual fault locator. For outside work, where much longer cable segments are the norm, an optical time-domain reflectometer (OTDR) with launch and receive cables and a fiber identifier are needed.

A fiber optic power meter is used in conjunction with a test source. Prices range from under $200 to over $1000 depending upon accuracy and extra features such as a high-capacity (1000 events) memory and ability to download to a computer. Also, a fiber optic meter plugs directly into any digital multimeter, so less equipment has to be purchased. The light source and patch cords are sold separately.

Outside fiber optic installation requires a much greater investment in equipment. A temperature-controlled van, bucket truck, trencher, and backhoe are essential. As for test equipment, the OTDR is needed. This $5000 instrument is used for verifying new installations and troubleshooting circuits that are performing poorly or not at all. The tool will pay for itself if it can locate a bad spot in a long underground cable so that extensive digging is not required.

The OTDR is a close relative to the time-domain reflectometer used for copper network cabling. The OTDR feeds light pulses into the optical fiber at one end. Instantly, it receives and reports any reflections that appear due to defects or anomalies anywhere along the line. These are similar to mismatched impedances in an electrical line and they can be due to design flaws or defects in the installation.

An OTDR has a sophisticated graphical interface as well as memory and computer compatibility, and a certain amount of training and on-the-job experience is necessary to accurately record and interpret the results. An OTDR is used at various stages of optical fiber manufacturing, warehousing, and shipping, and it is a good idea to take readings of the cable on the reel immediately prior to installation.

NEC Article 770 is devoted to optical fiber cabling, the concern not being electric shock or ignition of combustible material because that kind of energy is not available. Issues addressed by the Code are fire propagation (originating from a different source) and smoke generation. These two hazards are mitigated by defining abandoned optical fiber as installed optical fiber that is not terminated at equipment other than a connector and not identified for future use with a tag. A little further on, there is the requirement that the accessible portion of abandoned optical fiber cables be removed. Where cables are identified for future use with a tag, the tag must be of sufficient durability to withstand the environment involved. This definition in conjunction with the requirement is intended to solve the problem that exists in many buildings of a great proliferation of old, unused cabling above suspended ceilings, fastened to wall surfaces, and in hollow cavities. If there is a fire, hundreds of pounds of insulation will contribute to heavy, choking smoke and add to the available stock of fuel.

The same mandate applies to low-voltage wiring in general, but not to power and light conductors. In addition, it does not apply to wires or fibers that are not accessible. This includes optical fiber that is in a raceway. The reasoning is that if the fiber is in a raceway,

Figure 17-5 Optical fiber makes use of specialized terminations. (*Courtesy of Panduit.*)

it is not as likely to ignite. Furthermore, being in a raceway, the discontinued cable can serve as a pull rope for a future installation.

While this requirement goes far in making a better environment from the point of view of fire safety, there are situations where it will place a heavy burden on commercial building owners or tenants as well as the cabling technician. There can be a staggering amount of cable that has to be removed, all the while taking care that the existing network is not disrupted. Who is responsible for this work? The question does not seem to have a definitive answer at this time.

Another set of requirements intended to remedy the problem of spread of fire and smoke generation involves the hierarchy of cable types in various locations. Here again, optical fiber requirements parallel low-voltage cable requirements (see Fig. 17-5). We discussed them earlier. To review, plenum, riser, and general usage applications each take progressively less expensive cable types, which can be identified by the letter designations ONP, OFCP, OFNR, OFCR, OFNG, OFCG, OFN, and OFC. The cables higher on the list, intended for more sensitive locations, can be substituted for those farther down the list. For example, if you had a part reel of plenum-rated optical fiber, you could use it in a riser application.

The foregoing survey of troubleshooting and repair techniques for commercial electrical equipment has been intended to provide an overview. Complete coverage of this complex topic would require more volumes than we could imagine, but for now this ought to suffice to open some doors. Great care must be exercised when working with electricity, but this should not put a halt to our efforts to make use of this great force.

I believe that working with electrical equipment leads us into some profound areas. Phenomena encountered herein are based on elemental particles, and they are wondrous to contemplate. We can look closely at nature and marvel at the complex yet simple order of things. Our ancestors spoke of the music of the spheres. Perhaps it is fitting that many musicians (Phil Glass, Herbie Hancock, and Elvis Presley) were first electricians. The human soul may be more difficult to repair than a refrigerator, but we have to start somewhere.

Index